Worlds in the Sky

William Sheehan

Worlds in the Sky

PLANETARY DISCOVERY FROM

EARLIEST TIMES ✳ THROUGH

✳ VOYAGER AND MAGELLAN

The University of Arizona Press
TUCSON & LONDON

The University of Arizona Press
Copyright © 1992

The Arizona Board of Regents
All rights reserved

⊛ This book is printed on acid-free, archival-quality paper.
Manufactured in the United States of America

97   96   95   94   93   6   5   4   3   2

Library of Congress Cataloging-in-Publication Data
Sheehan, William, 1954–
    Worlds in the sky : planetary discovery from earliest times
    through Voyager and Magellan / William Sheehan.
        p. cm.
    Includes bibliographical references (p. 223) and index.
    ISBN 0-8165-1290-6 (cloth)
    ISBN 0-8165-1308-2 (paperback)
    1. Planets.   2. Astronomy—History.
    3. Planets—Exploration.
I. Title.
QB601.S543   1992   91-39398
523.2—dc20   CIP

British Cataloguing-in-Publication Data
A catalogue record for this book is available from
the British Library.

To Richard Baum and Harold Hill

CONTENTS

FIGURES

Note: Telescopic observations are reproduced here with north at the bottom. Spacecraft images are shown with north at the top.

ACKNOWLEDGMENTS

This book is intended as an introduction to planetary astronomy. There are many such books, as the subject is an immensely attractive one, given the incredible explosion of knowledge about the planets and their moons that has taken place over the past decade or so. It is therefore only reasonable to ask, why yet another?

What distinguishes this book is that it is not content to treat only the recent spacecraft findings; rather, its subject is nothing less than the whole history of humankind's fascination with the Moon and the planets, and that fascination began in the naked-eye era. In this book I give special prominence to visual telescopic work, which I believe should not be forgotten even with all that the spacecraft have achieved. The actual sight of lunar mountains or of planetary markings "imparts a sense of satisfaction which the armchair astronomer neither knows nor deserves to know," as H. P. Wilkins, one of the old observers, put it. That being the case, it is an undeniable fact that the great observers of the past are better guides to what can actually be seen in modest instruments than the Voyager spacecraft cameras. This, then, points to a major purpose of the book: to place the heroic work of the past in the context of recent discoveries.

Many people have helped to make this book a reality. Richard Baum, director of the Terrestrial Planets Section of the British Astronomical Association, provided great impetus to its creation by sharing with me his enthusiasm and vast expertise in planetary astronomy. He read the manuscript attentively and very generously offered to prepare some of his exquisite planetary drawings specially for this work. Harold Hill has similarly enriched the chapter on the Moon with some of his beautiful pen-and-ink stipple and ink-wash drawings. The work of these exceptionally gifted observer-artists has allowed me to avoid relying exclusively on images that have, I fear, become hackneyed through overexposure. Moreover, they show what can yet be accomplished using "old-fashioned" visual methods.

Others who graciously supplied illustrations were Dr. Audouin Dollfus of the Meudon and Pic du Midi observatories in France; Dr. Donald E. Osterbrock of the Lick Observatory and Dorothy Schaumberg of the Mary Lea Shane Archives of the Lick Observatory; and Dieter Gerdes of the Heimatverein Museum in Lilienthal, Germany. Also, I owe a debt of gratitude to Melanie Magisos of the Lunar and Planetary Laboratory of the University of Arizona in Tucson, whose help in obtaining spacecraft images was indispensable.

Barbara Beatty was a great source of encouragement in the early stages of the project, and her successor at the University of Arizona Press, Jennifer Shopland, carried it through with aplomb. The entire staff at the Press, especially my copyeditor, Alan Schroder, have been a joy to work with. Michael Conley deserves great thanks for his considerable assistance in getting the manuscript ready for publication. But my deepest gratitude is to my wife, Deborah, and my two young sons, Brendan and Ryan, who patiently endured my absences when I was poring over these pages. Each in his or her own way nurtured the project at crucial times and, perhaps equally important, helped me to keep a healthy perspective on it.

WILLIAM SHEEHAN
St. Paul, Minnesota
December 1991

INTRODUCTION

This book seeks to tell the story of humanity's quest for the planets. It is a story that can be divided into three parts. The first is by far the longest in years and the shortest in results. It begins with the Chaldeans, or whoever the still-earlier stargazers may have been who discovered that five "stars" do not keep to the same place as do all the rest but instead are wanderers among them—planets, as the Greeks called them.

The early stargazers had only their eyes to help them, and the main problem they faced was to account for the motion of the planets. The best naked-eye observations that are humanly possible are accurate to several minutes of arc, an accuracy routinely achieved by the last of the great naked-eye observers of the planets, the Danish nobleman Tycho Brahe, who died in 1601. From Tycho's observations, his mathematically gifted assistant Johannes Kepler was able to determine that the paths in which the planets travel through space are shaped like ellipses. This was the greatest triumph of the naked-eye era of astronomy.

Kepler announced this result in 1609, and in that same year a professor of mathematics at Padua, Galileo Galilei, and several others, including Thomas Harriot of England and Simon Marius of Germany, turned the first telescopes toward the sky and initiated the second period of planetary discovery.

The Moon, which reveals only a few bright and dark splotches to the naked eye, shows in the telescope a terrain with high and low levels, whose true character is betrayed by the changing appearance of the features under different conditions of illumination. The shadows cast by the higher elevations crawl across the surface and cast into sharp relief even modest irregularities. Similarly, the planets, rather than being merely brilliant dots of light, are magnified into disks with their own characteristic markings—other worlds and perhaps, the early observers surmised, other Earths.

Three centuries of careful work by astronomers peering through telescopes and through the troublesome medium of the Earth's atmosphere sketched in a number of the details. By 1960 it seemed to be firmly established that Mercury, the innermost planet, always kept the same face turned toward the Sun, just as the Moon does toward the Earth. The surface of Venus, hidden as it was beneath opaque clouds, was completely unknown, and deserts and jungles there seemed equally tenable. It was still widely believed that most of the craters of the Moon had been formed by volcanic processes; that Mars supported primitive life-forms; that Jupiter's Great Red Spot was some kind of solid object floating in the planet's atmosphere, rather like an egg floating in a glass of saltwater. As far as anyone knew, Saturn's rings, the aesthetic glory of the Solar System, were unique. Not a single one of these ideas has stood the test of time. Indeed, some of them now seem rather quaint.

By 1960 the first spacecraft were being launched toward the Moon and the nearby planets, and with this we come to the third period of planetary exploration, in the midst of which we find ourselves today. This third period is easily the most remarkable. The increase of our knowledge has been commensurate with the ever greater detail of the views we have obtained, producing an advance every bit as great as that which the telescope brought over the naked eye. What we had before were mere chapbooks about the Moon and planets. Now there is enough information to fill an encyclopedia of many volumes. It is fair to say that planetary science is no longer an academic discipline remote from everyday life. On the contrary, we are beginning to understand the balance of factors that has allowed the environment of the Earth to sustain life. Elsewhere—on Venus, for instance—we can see the result when the balance tips in a different direction. We still have much to learn, but it is already clear that what we learn will prove indispensable to our very survival.

My goal in this book is to describe not only where we are now in our understanding of the planets but also something of how we got here. My justification in taking this approach is the fact that knowledge is a product of history and thus is subject to the influence of the people who obtained it. Though I have tried to make this book as complete and up-to-date as possible with regard to the recent spacecraft discoveries, it is important also to remember the travails of those who have gone before. The story of their efforts is interesting in its own right, and it is instructive as well in showing the roundabout way along paths of error by which science attains understanding.

Worlds in the Sky

CHAPTER ONE

Wandering Stars

Of the thousands of stars visible to the naked eye, only five seem to change their places with respect to each other and the rest of the stars. These are the planets, which from the earliest times excited the greatest wonder. Their motions are not completely regular; to the contrary, at times they move more swiftly, at times more slowly, and at still other times they stand still—at their so-called stationary points—and then move backward, or retrograde, for a period before again coming to a standstill and then resuming their usual motion. "Such irregularity of motion," wrote Geminus, a Greek astronomer who lived at Rhodes in the first century B.C., "would not even be expected of a decent and orderly man in his journeys." How much less, then, of the imperishable stars?

Of the planets, two, Mercury and Venus, lie closer to the Sun than the Earth and always remain close to the Sun in the sky, Mercury never departing from its side by more than 28°, and Venus by no more than 47°. They may pass behind the Sun (into superior conjunction) or between the Earth and the Sun (into inferior conjunction). If they could be observed when they were between the Earth and the Sun, the inferior planets would be found to be moving retrograde, but of course they are then hidden by the solar glare.

The outer planets—Mars, Jupiter, and Saturn—move retrograde at about the time they are opposite to the Sun in the sky, or at opposition, a situation which, of course, is never possible for a planet that is closer to the Sun than is the Earth. Far from being hidden, the outer planets are at their brightest when in opposition, and thus the changes in direction that they undergo are readily noticed and came to the attention of early stargazers. This was especially so in the case of Mars.

The first attempt to account for the motions of the planets mathematically was made by the Greeks. On religious and philosophical grounds, the Greeks believed the heavens to be perfect and immutable. This meant that underlying the apparent irregularities of the planets' motions, the true motions had to be completely uniform and circular (the circle was regarded as the most perfect form). Most of the Greeks also assumed that the Earth was the center of the universe. The question therefore became, How can the observed motions be explained, taking the Earth as the center and allowing nothing but uniform circular motion?

Various attempts were made to solve the problem. There was, for instance, the ingenious system of homocentric spheres developed by Eudoxus of Cnidus, a student of Plato who died in 355 B.C. Eudoxus tried to reproduce the apparent path of each planet, particularly the retrograde movements, by summing the motions of several nested spheres. Though the spheres had different axes of rotation, all were centered on the Earth. As far as we know, Eudoxus regarded his spheres as a mere mathematical device and did not attribute to them a real existence, though Aristotle, the most influential philosopher of antiquity, later did so. Be that as it may, the Eudoxan scheme was obsolete even before Aristotle died in 322 B.C. A fatal objection to it was that, since the spheres were all centered on the Earth, the variation in the planets' brightness (in the case of Mars amounting to some fiftyfold) could not be explained.

This leads us to the ideas of Aristarchus of Samos, who from the modern perspective was the greatest of the Greek astronomers. Aristarchus realized that some other point besides the Earth had to be the center of the planetary motions and boldly chose the Sun, demoting the Earth to the status of an ordinary planet orbiting around it. With this masterly stroke, the retrograde motions of the planets were revealed to be reflections of the Earth's orbital motion. They are apparent displacements produced by parallax, because the planets are viewed from different points as the Earth traverses its path around the Sun.

How did Aristarchus ever arrive at his heliocentric system? One possibility is suggested by the only work of his which has survived—his treatise "On the Distances of the Sun and Moon." Here Aristarchus proposes that by measuring the Sun–Earth–Moon angle around the time the Moon's phase is exactly half, one can work out the relative distances of the Sun and the Moon. This angle is 89.8°, though Aristarchus used 87°, which indicates he can hardly have actually measured it, for even with a simple gnomon, or stick in the ground, one can do much better than that. Yet though off by a factor of twenty, Aristarchus's estimate was good enough to make him realize that the Sun had to be an enormous body, many times larger than the Earth, and this may well have been what started him thinking in heliocentric terms.

Bold as it was, Aristarchus's heliocentric system remained an over-simplification, inasmuch as it still had the planets moving in circular orbits—a not very accurate approximation. Thus his theory did not exactly agree with what was observed, and the later Greeks cannot really be blamed for inventing less jolting ways of "saving the phenomena," as they described the problem of constructing a geometrical device that would represent the observed motions in the sky.

Translating Aristarchus's heliocentric system into a geocentric scheme gives rise to a construction in which a planet moves in a little circle (known as the epicycle), which in turn moves about a larger circle (the deferent) centered on the Earth. This scheme is mathematically equivalent to Aristarchus's heliocentric system, but it had the decided advantage (from the ancients' point of view) of retaining the Earth as the center.

Claudius Ptolemy, a Greek astronomer living in Alexandria, Egypt, in the second century A.D., brought this system of epicycles and deferents to its fullest elaboration. Thus it is always referred to as the Ptolemaic system, though in fact Ptolemy did not invent it.

In order to account for small irregularities in the observed motions of the planets that were discovered after Aristarchus's time, Ptolemy found it useful to place his deferents slightly off-center from the Earth. In addition, he made the motion of each epicycle around the deferent take place uniformly not with respect to the center but with respect to another locus, the equant, situated so as to make the center of the deferent lie exactly at the midpoint between the Earth and the equant itself. This system accounted for the observed planetary motions tolerably well. However, even Ptolemy could hardly have regarded the device as literally true—after all, it involved a tacit abandonment of the principle of uniform circular motion, which for the Greeks was nothing less than heresy.

Though Ptolemy's system was to be the last word in astronomy for over a thousand years, his reputation has suffered greatly in later centuries. Merely to mention his name is to conjure up visions of all those epicy-cles—some eighty in all. Indeed, the Ptolemaic system has become almost synonymous with entrenched scientific error. One recalls the famous remark of Alfonso X of Castile, whose tables of planetary motion based on Ptolemy's theory were to remain the standard for two centuries after Alfonso's death in A.D. 1284. Ptolemy's system, said Alfonso, was "a crank machine; it is a pity the Creator had not taken advice." Yet if a theory is to be judged solely by the accuracy with which it accounts for the observations, then Ptolemy's theory was clearly a success; it was accurate to within the tolerance of the measuring instruments of his time.

Ptolemy's system was the swan song of Greek science. With him its work was to all intents and purposes finished. Indeed, the whole classical world would soon crumble. In A.D. 389 the great library of Alexandria was destroyed by order of Theophilus, the Christian bishop of Alexandria, who was fanatically opposed to anything pagan. This is as convenient a date as any to take as the beginning of the Dark Ages, and astronomically speaking they were indeed dark. Such basic ideas as the sphericity of the Earth, already well established in the time of Aristotle, were condemned by the Church Fathers on the grounds that they seemed to be inconsistent with Holy Scripture. Meanwhile, the learning of the Greeks was generally forgotten. Credit for preserving what little of it survived belongs to the Arabs, whose holy book, the Koran, contained no opinions about the true construction of the world. Indeed, some of the Arabs were skilled observers in their own right—a fact recalled in the many Arabic names still used for the stars.

Eventually the passion for learning revived in Europe. Artists pondered the ruins of ancient buildings, and scholars rediscovered the long-neglected texts. Filippo Brunelleschi, for example, studied the Pantheon in Rome in preparing his plans for the cupola of the church of Santa Maria del Fiore in Florence, which called for a dome far greater than that of the Pantheon. "I wonder," he wrote in 1417, "if even the ancients ever raised a vault as daunting as this will be." In Ermland, now a part of Poland, an astronomer was born in 1473 who would undertake the task of redesigning the vault of the heavens.

As a young man, Nicolaus Copernicus prepared for a career in the Church, and when he was twenty-five the bishop of Ermland, who was his uncle, arranged for him a sinecure in the Cathedral of Frauenburg, though he was some years in assuming it. Copernicus also studied medicine and served as his uncle's personal physician for a time. Astronomy was his true passion, however, and in the course of his studies he became dissatisfied with Ptolemy's theory and began to wonder whether a better model of the heavens might not be constructed. One of his chief objections to Ptolemy was that he had, as we have seen, abandoned the strict principle of uniform circular motion. Copernicus found among the classical texts a few references to other ideas about how the universe might operate, including those of Aristarchus, and with this encouragement he had worked out his own heliocentric theory by about 1512.

He found that having the Earth move around the Sun eliminated Ptolemy's principal epicycle for each planet. This is hardly surprising, since that epicycle merely reflected the Earth's own motion around the Sun. However, he remained committed to circular orbits for the planets, and, needless to say, this was no more successful as a basis for detailed calculations than it had been when Aristarchus had proposed it seventeen centuries before. Indeed, it was considerably less successful than Ptolemy's theory. Yet, whereas Aristarchus's successors had abandoned the heliocentric system at precisely this point, Copernicus stuck with it. One of the reasons for this, I suspect, was that Copernicus was familiar with the principle of scientific perspective, which Brunelleschi and others had worked out. As William Ivins, Jr., writes in *Art and Geometry,*

> Technically, perspective is the central projection of a three-dimensional space upon a plane. Untechnically, it is the way of making a picture on a flat surface in such a manner that the various objects represented in it appear to have the same sizes, shapes, and positions, *relatively to each other,* that the actual objects as located in actual space would have if seen by the beholder from a single determined point of view.

The Greeks, for all their ingenuity, had never solved this problem, but Copernicus must have realized that, conceptually, it was the reciprocal of the problem of planetary motion. Scientific perspective involves projecting three dimensions onto a plane; the planetary problem involves deducing from the two-dimensional picture of the planets' apparent motion their actual motion in space relative to the moving Earth. Copernicus wrote of the retrograde motion of the outer planets:

> This happens by reason of the motion, not of the planet, but of the Earth changing its position in its orbit. For since the Earth moves more rapidly than the planet, the line of sight directed toward the firmament regresses, and the Earth more than neutralizes the motion of the planet, that is, when it comes between the Sun and the planet at the planet's evening rising [opposition]. . . . When the line of sight is moving in the direction opposite to that of the planets and at an equal rate, the planets appear to be stationary, since the opposed motions neutralize one another.

Since the Greeks did not possess the secret of scientific perspective, this kind of argument would not have occurred to them. To a Renaissance man like Copernicus, it would have seemed obvious.

In a geocentric system, there is no way to deduce the relative positions of the planets in their orbits. Thus the early Greeks had arbitrarily adopted the order, moving outward from the Earth, of Moon, Sun, Venus, Mercury, Mars, Jupiter, Saturn. Later this scheme was changed to Moon, Mercury, Venus, Sun, Mars, Jupiter, Saturn. This was the order given by Ptolemy, but there was never any way, within the Ptolemaic system, of proving that this was the correct arrangement. In contrast, Copernicus's arrangement of the planets follows the modern one of Mercury, Venus, Earth, Mars, Jupiter, and Saturn, and it derived directly from his theory.

Though he replaced the Earth with the Sun as the center of the universe, in other respects Copernicus merely rehashed Ptolemaic astronomy. For instance, when he tried to make detailed calculations, he saw no alternative but to resurrect the whole machinery of epicycles, and in the end his system was nearly as complex as the Ptolemaic system it was intended to replace, requiring no less than forty circles in all, only a modest improvement over Ptolemy's eighty. His final theory was presented in his great book *De Revolutionibus Obrium Caelestium* (On the revolutions of the heavenly orbs), which was published in 1543. It is said that the first printed copies reached Copernicus when he was on his deathbed. Though he made mistakes, he had taken the decisive step, and henceforth progress was rapid.

The next step was to obtain better observations of the planetary motions themselves, and this was to be the main work of Tycho Brahe, a Danish nobleman born three years after Copernicus's death. Tycho made every effort to obtain the best instruments, and his observations were remarkably accurate, given that they all had to be made with the naked eye, as the telescope had not yet been invented. Copernicus had remarked once that he would have been "as happy as Pythagoras is said to have been on discovering his theorem could I be certain of my observations to within ten minutes of arc." Tycho generally achieved an accuracy to within two or three minutes.

Following Tycho's death in 1601, his treasury of observations fell into the hands of his young assistant, Johannes Kepler. Kepler was born in 1571 at Weil, a village in Württemberg, and by temperament he was inclined to mysticism. Some of his arguments sound absolutely absurd today, but he proved to be the perfect man to make use of Tycho's observations. He was a mathematician of great skill and, moreover, a confirmed Copernican, unlike Tycho himself, who had adopted a compromise position in which the planets revolved about the Sun while the Sun and Moon in turn revolved about the Earth (the Tychonic system). Kepler centered his attention on Mars and from Tycho's observations attempted to trace out the shape of the planet's orbit in space. It was obvious that it was shaped somewhat like an egg, but for a long time the exact solution eluded him. Finally he received his illumination. "I awoke as from a sleep," he wrote. "A new light broke on me." What he had discovered was his first law of planetary motion, published in 1609:

> The planets move around the Sun in elliptical paths, with the Sun at one focus and the other focus empty.

(For the sake of completeness, his two other laws are "The radius vector, that is, the line connecting the planet to the Sun, sweeps out equal areas in equal times," and "The squares of the periods in which the planets describe their orbits are proportional to the cubes of their mean distances from the Sun.")

Pondering the fact that the angular velocity of the planets decreases as their distance from the Sun increases, Kepler had speculated early in his career that "Either the *animae motrices* [moving souls] are feebler as they are more distant from the Sun or there is only one *anima motrix* in the center of all the orbits, that is, the Sun." Later he substituted the word *vis* (force) for *anima* (soul), and taking his hint from the laws of magnetism recently published by England's William Gilbert, suggested that "the celestial machine is . . . a clockwork, . . . inasmuch as all the variety of motions are carried out by means of a single very simple magnetic force of the body, just as in a clock all motions arise from a very simple weight."

Kepler's theory was that the Sun's magnetic force carried the planets around as the Sun rotated. Soon after he made this suggestion, sunspots were discovered by Fabricius, Scheiner, and Galileo, establishing the fact of the Sun's rotation, and Kepler regarded his view as confirmed. It was still far from clear, however, why the paths of the planets should be elliptical rather than circular. Kepler proposed that every planet had a magnetic axis whose direction was fixed in space. The poles of each planet were taken to have opposite magnetic polarity, one being "friendly," the other "antagonistic" to the Sun. Depending on which pole was turned toward the Sun at a given time, the planet was either attracted or repelled and so was deflected from a circular path. Obviously Kepler was groping toward a dynamic explanation of planetary motion, but he never completely solved the problem. That was to be the majestic achievement of

Isaac Newton, who showed that planetary motion could be accounted for by assuming only an attractive force—gravitation.

Newton was born at Woolsthorpe, in Lincolnshire, in 1642. When he was nineteen he went up to Trinity College, Cambridge, but his education was interrupted by the Great Plague of 1665. The University shut down and Newton returned to his mother's farm at Woolsthorpe. There he began his research into gravitation—according to a story he himself told in his later years, the chain of his reasoning began with the fall of an apple in his mother's garden. "I began to think of gravity extending to the orb of the Moon, . . . compared the force requisite to keep the Moon in her orb with the force of gravity at the surface of the earth and found them to answer pretty nearly." Newton's idea was that the gravitational force was proportional to the inverse square of the distance, but he did not publish it at the time.

After the plague subsided, Newton returned to Cambridge, becoming Lucasian professor of mathematics, but for a number of years his work on gravitation was set aside. Only when Robert Hooke, a leading member of the Royal Society whom Newton detested, came to the same conclusion about the inverse square law and even gave strong hints that he had priority in the matter was Newton at last roused to action. Encouraged by Edmund Halley, he wrote out his mathematical tour de force, *Philosophiae Naturalis Principia Mathematica* (Mathematical principles of natural philosophy). It was fifteen months in the writing and was published in 1687, Halley generously assuming the cost of publication.

The *Principia* is one of the immortal works of science. In it Newton showed that the tides in the Earth's oceans are due to the gravitational attraction of the Moon, and that the precession of the equinoxes (the slow drift of the point in the sky where the Sun is located as it crosses the celestial equator in the spring or autumn) is another effect of the Moon's pull on the Earth, which produces a slow wobbling of the Earth's axis. He calculated the Earth's oblate shape, derived the Keplerian ellipses that the planets follow around the Sun, and made a heroic start on the difficult problem of the perturbations in the motion of one planet caused by the gravitational attraction of another. In the case of the Moon, which travels around the Earth but which is at the same time disturbed by the Sun, the problem of calculating the resultant motion is especially difficult. It is the classic example of what is known in gravitational astronomy as the three-body problem. Newton's goal was ambitious: to produce a lunar theory accurate to within 2 or 3 minutes of arc. Accurate observations were needed, and at the time the best available were those of John Flamsteed, who had been Astronomer Royal at the Greenwich Observatory ever since its founding in 1675. Flamsteed supplied some observations, but Newton was a difficult man to satisfy, and a quarrel inevitably broke out. By the time Flamsteed died in 1719, they had become implacable enemies. The lunar theory became one of Newton's bitterest disappointments (the problem of the Moon was, he once said, the only one that ever made his head ache). His final theory was accurate only to 10 minutes of arc, and not until the next century, with the work of Euler, Clairaut, d'Alembert, and Laplace, was Newton's goal of achieving accuracy to within 2 or 3 minutes finally achieved.

Newton himself believed that the orderly arrangement of the Solar System showed it to be the result of a "cause . . . not blind and fortuitous, but very well skilled in mechanics and geometry." However, as we now know, the arrangement owes at least as much to chance as to design. The Solar System began with a swirling nebula. As it rotated, the nebula formed a flattened disk. Matter accumulating at the center became the Sun, while the disk broke into a series of rings from which, grain by grain, small planetesimals formed. These accreted into planets. The inner planets are small and rocky, while the outer planets are giants made up chiefly of gas.

During the late stage of the accretion process, the planets were subjected to a violent bombardment of their surfaces, the scars of which are visible in the craters of the Moon and other bodies throughout the Solar System. In a few cases the collisions were on a stupendous scale. For instance, a Mars-sized planet seems to have grazed the proto-Earth, giving rise to the Earth-Moon system we have now, while other planetary collisions may explain Mercury's unexpected iron core, Venus's slow rotation in the "wrong" direction, and Uranus's being tilted onto its side. This is a far cry from Newton's "cause . . . well skilled in mechanics and geometry." It is instead a process that smacks much more of the "blind and fortuitous."

With Newton, the revolution that had begun with Copernicus was completed. Some of the details of the motion of the Moon and the planets still had to be worked out, but the main results were firmly in place. There were, however, other worlds to conquer. With the invention of the telescope, astronomers could ponder not only how the planets moved but also what they were, and indeed their surfaces soon revealed tantalizing details—what seemed to be far-off continents, oceans, and clouds.

CHAPTER ✳ TWO

Through the Telescope

W̲hen one thinks of the invention of the tele-
scope, one thinks, of course, of Galileo—even though he neither invented
it nor was he the first to turn it toward the sky. We now know, for instance,
that an Englishman, Thomas Harriot, sometime friend of Sir Walter
Raleigh, made the first telescopic observations of the Moon several
months before Galileo, and that others were active at about the same time.
Yet though Galileo was not alone as a telescopic pioneer, he was far in
advance of the others in the perfection of his telescopes, the diligence of his
observations, and the soundness of his interpretations, so that most of the
credit is deservedly his.

Galileo was born in 1564, the same year as Shakespeare. His father was
Vincenzo Galilei, a musician, who hoped his son would one day become a
physician. At seventeen, Galileo was sent to the University of Pisa to begin
the study of medicine, but he was not especially drawn to the subject, and
besides, he could not keep up with the tuition, so in 1585 he left without
completing his work toward the degree.

Galileo's real interest was mechanics. Even before leaving the University
of Pisa he had discovered the fundamental law of the pendulum: a pendulum of a given length swings with the same period regardless of the
amplitude. Soon afterward he produced a hydrostatic balance for determining the density of materials. The university authorities were suitably
impressed, and in 1589 he was recalled to Pisa as an instructor of mathematics. Three years later, Galileo resigned in order to take a more lucrative
position at Padua, in the Republic of Venice, and there he became con-
verted to the Copernican theory. As he wrote to Kepler in 1597, "Like you,

. . . I have written up many reasons and refutations on the subject, but I have not dared . . . to bring them into the open. . . . I would dare publish my thoughts if there were many like you; but, since there are not, I shall forbear." As Galileo knew, there was as yet no incontrovertible proof of the Copernican theory. But then came the invention of the telescope by a spectacle maker in Flanders in 1608 by means of which, as Galileo himself wrote, "visible objects, though very distant from the eye of the observer, were distinctly seen as if nearby." Galileo heard of the invention during the summer of 1609 and was able to work out the principles for himself, arranging two lenses, one convex and one concave, in a lead tube to produce a crude telescope with a magnification of three. Before long, he had an instrument magnifying eight or ten times, which he was showing off to the senators and other notables from atop the highest campaniles in Venice.

Galileo seems to have been preoccupied for a time with the instrument's commercial possibilities; not until November 1609 did he actually turn a telescope toward the heavens. His first object, naturally enough, was the Moon, which he recognized "is not smooth, uniform, and precisely spherical, as a great number of philosophers believe it (and the other heavenly bodies) to be, but is uneven, rough, and full of cavities and prominences, being not unlike the face of the earth, relieved by chains of mountains and deep valleys." Indeed, he even managed to measure some of the mountains, finding them comparable to the Earth's mountains in scale. By January 1610 he was observing the four large satellites of Jupiter—"four wandering stars," as he put it, "not known or observed by any man before us." Shrewdly, he placed the satellites and himself under the protection of the powerful Florentine House of Medici, naming the new Jovians the "Medicean Stars." He described his observations in a lively little book, *Sidereus Nuncius* (the Starry Messenger), which he wrote at lightning speed and published in Venice in March 1610.

His book, with its dedication to Cosimo de' Medici II, helped Galileo obtain the position of "Chief Mathematician and Philosopher" to the Medici in Florence. He arrived in Florence in September 1610 and at once began making some of his most crucial observations—those on the phases of Venus. According to Ptolemy, Venus ought never to show a full disk, or even a half, whereas Copernicus had predicted that the planet would pass through the entire cycle of phases (all of this assuming, of course, that Venus shone by reflected light rather than by light of its own and that it did not keep to a course lying always beyond the Sun). From late summer to early November 1610, Venus would have been too small in Galileo's telescope to show anything of interest. Only at the end of December did the planet clearly show a crescent shape, proving to Galileo that the Ptolemaic explanation could not fit the facts and that the Copernican system had to be the correct one.

The Tychonic theory made the same prediction as the Copernican, so geocentrists could still fly there for refuge, but at first Galileo carried all before him. In Rome, Pope Paul V received him with enthusiasm, and the leading Jesuit astronomers relinquished their initial skepticism about his discoveries after looking through his telescope. Back in Florence, Galileo made out the "triple form" of Saturn, which greatly puzzled him, and the sunspots, which he observed independently of but somewhat later than Johannes Fabricius in Holland and Christoph Scheiner in Germany. Scheiner was a Jesuit, and in order to avoid attributing blemishes to the Sun itself, he suggested that the spots were small planets seen in front of the solar disk. Galileo refuted Scheiner's position in his *Letters on Sunspots* of 1613, pointing out that the spots had to be something actually on the surface of the Sun itself, as they could be shown to rotate with it. His own view was that they were on the order of "vapors, or exhalations, . . . or clouds, or fumes." Never one to be modest, he claimed for himself full credit for the discovery of sunspots: "It was granted to me alone to discover all the new phenomena in the sky and nothing to anybody else," he wrote. "This is the truth which neither malice nor envy can suppress." In fact this was untrue, and the priority dispute made Scheiner his enemy for life.

Soon afterward Galileo was drawn into other, more dangerous controversies. At the end of 1613 a priest attacked him from the pulpit, claiming that the Copernican system was "hostile to divine Scripture." Galileo wrote out a long response, which said in part:

> Who should set a limit to the mind of man? Who would dare assert that we know all there is to be known? Therefore, it would be well not to burden the articles concerning salvation and the establishment of the Faith—against which there is no danger that valid contradiction ever may arise—with official interpretations beyond need; all the more so, when the request comes from people of whom it is permitted to doubt that they speak under heavenly inspiration, whereas we see most clearly that they are wholly devoid of that understanding which would be wholly necessary, I will not say to refute, but first of all to grasp the demonstrations offered by science.

Wise advice, perhaps, but the theologians were bound to resent this incursion into their domain by a mere professor of mathematics. A file on

Galileo was opened in the Holy Office of the Inquisition, and in 1616 *De Revolutionibus* was put on the Index of Forbidden Books and Galileo was warned to refrain from teaching its doctrines.

He did so for a time. Then in 1623 Cardinal Maffeo Barberini, who had been Galileo's friend in earlier days and who had opposed the ban of 1616, became Pope Urban VIII. Thinking that the intellectual climate had changed, Galileo was emboldened to write his *Dialogue Concerning the Two Chief World Systems—Ptolemaic and Copernican*. The book was unabashedly partial to Copernicus. However, Galileo had miscalculated about Barberini. As pope he was less open-minded than he had been as cardinal, and eventually Galileo was summoned to Rome to face the Inquisition. On his knees in the Convent of Minerva in Rome on June 22, 1633, he was forced to "abjure, curse, and detest" his errors.

Galileo was now an old man, and his eyesight, which had made him the first man to behold so many wonders, was beginning to fail. By the end of his life, he was totally blind. He was placed by the Inquisition under house arrest at his villa in Arcetri, near Florence, where he died in 1642 (the year of Newton's birth). The Catholic church's ban against Copernicanism stifled the pursuit of science in Italy. *De Revolutionibus* was not removed from the Index until 1835. Henceforth the initiative passed to more liberal countries.

It is worth remembering that all of Galileo's telescopic discoveries were made with instruments that had a simple convex lens as the objective and a concave lens as the eyepiece, and that even his best telescope magnified only about thirty times. The field of view was very small; moreover, the lenses were not accurately figured, and in order to improve the image Galileo had to introduce cardboard rings in front of the object glass so that the light would pass only through the part of the lens where the curvature was most uniform.

The discovery of sunspots was the last major discovery of the initial phase of telescopic astronomy. Galileo had brought the simple "Dutch" type of telescope as far as he could; the next wave of discoveries awaited further improvements in the instruments. One such improvement was the replacement of the Dutch telescope, with its concave eyepiece, by the "astronomical telescope," which used a convex lens as the eyepiece. Still, there were serious problems. The early refractors suffered horribly from their tendency to generate fringes of false color around bright objects. Venus, for example, flamed with prismatic splendor. The cause of this false color, or chromatic aberration, was unknown at the time, but as Isaac Newton later showed, white light is a mixture of all the colors of the rainbow, and on passing through a simple lens the colors are bent by different amounts, blue more than red. Thus the different colors form images subtending different angles when viewed through an eyepiece, with the result that the object appears to be surrounded by a halo of vivid colors, which makes it impossible to bring it into sharp focus.

The early observers naturally wanted to increase the aperture and magnifying power of their telescopes. Unfortunately, the larger the aperture, the worse was the chromatic aberration. Eventually astronomers found that by making the curvature of the lenses shallower and the focal length longer, the effect of chromatic aberration was reduced, though each increase of aperture meant an increase by the square of the focal length. In other words, doubling the aperture meant increasing the focal length by a factor of four, tripling the aperture meant increasing the focal length by a factor of nine, and so on.

One of the first to build long telescopes was Christiaan Huygens of Holland, who also found better ways of figuring lenses and devised the first compound eyepiece, still known as the Huygenian. With a 2-inch telescope of 10 1/2-foot focal length, magnifying 50×, he discovered Saturn's largest moon in 1655 and soon afterward, using a 23-foot telescope, recognized that the planet was surrounded by a ring. Inspired by Huygens's discoveries, Johannes Hevelius, a brewer and later a city councilor of Danzig (now Gdansk), who was one of the most enthusiastic instrument makers and observers of the generation after Galileo, built telescopes with focal lengths of 60, 70, and finally 150 feet, though the last was so unwieldy that its tube shook like a reed in the wind and could rarely be used to advantage. It was destroyed in the great Danzig fire of 1679.

Because of the unwieldiness of the very long tubes used by Hevelius, Huygens turned his attention during the 1680s to the tubeless "aerial telescope," in which the object glass was fixed to the top of a tall mast, and the observer sighted along guy wires, which could be used to point the object glass in any desired direction. The observer simply held the eyepiece by hand. It goes without saying that such instruments were difficult to use, but they were a definite improvement on instruments of the Hevelian type. Huygens erected one of 123-foot focal length at his country estate near The Hague and followed this with one of 210 feet. He died in 1695.

Huygens's chief rival in planetary discovery was Giovanni Domenico Cassini, who was born in 1625 at Perinaldo, near Nice, in what was then the Duchy of Savoy. As a young professor at the University of Bologna, Cassini won fame for his accurate measurements of the rotations of Mars

FIGURE 2.1. Christiaan Huygens (1629–1695)
(Courtesy of Yerkes Observatory)

and Jupiter, which were made using telescopes built by the Roman instrument maker Giuseppe Campani. In 1669 Louis XIV invited him to Paris to become director of his new national observatory. (Huygens, incidentally, was also living in Paris at the time. He had come at Louis' invitation in 1666 and remained at the Bibliothèque du Roi for fifteen years, until increasing intolerance for Protestants in Catholic France caused him to return to Holland.)

On arriving in Paris, Cassini found that detailed plans for the observatory had already been drawn up by Claude Perrault, the architect responsible for designing the new façade at the Louvre. Cassini, with an astronomer's rather than an architect's eye, strongly objected to the plans. Eventually a meeting was arranged between Cassini and Perrault, with the king and his minister of finance, Jean-Baptiste Colbert, also on hand. Cassini's great-grandson Jean Dominique Cassini IV later described the meeting:

> Perrault eloquently defended his plan and architectural style with beautiful sentences. My great-grandfather spoke French very poorly, and in defending the cause of astronomy he shocked the ears of the King, Colbert, and Perrault to such a point that Perrault in the zeal of his defense said to the King: "Sire, this windbag doesn't know what

he is saying." My great-grandfather kept silent and did well. The King agreed with Perrault and did badly. The result is that the observatory has no common sense.

Fortunately Cassini proved to be a resourceful man. He simply set up the long telescopes he had brought with him from Italy in the courtyard outside the observatory and went to work. Between 1671 and 1684, Cassini's keen eye discerned the main division in Saturn's rings and four new Saturnian satellites, which Cassini named the "Louisian stars" in honor of his patron. The two faintest, later renamed Dione and Tethys, were glimpsed using Campani "aerial" telescopes of 100 and 136 feet—surely a remarkable feat.

The "aerial" telescopes of Huygens and Cassini were awkward by any standard, and in any case, by the end of the seventeenth century they had already attained the greatest possible length. "Seeing therefore the improvement of telescopes of given lengths by refractions is desperate," Isaac Newton wrote in his *Opticks* of 1707, "I contrived heretofore a perspective [i.e., a telescope] by reflexion, using instead of an object-glass a concave metal." Newton's idea, in other words, was to use, in place of a lens, a curved mirror (in those days made of speculum metal rather than aluminum-coated glass as today) to collect the light. The main mirror reflected the light back up the tube to a small, flat mirror set at an angle so as to redirect the beam through a hole in the side of the tube, where the image could be magnified by an eyepiece. Newton's first reflector, built with his own hands, had a mirror of 1-inch aperture and was presented to the Royal Society of London in 1672. There have been many variations on Newton's basic design. In the Cassegrain reflector, for example, the main mirror has a hole in it, and the light is reflected directly back to the eyepiece rather than being deflected to the side. But these are details. The important point is that the reflector avoids the problem of chromatic aberration altogether.

For various reasons, the earliest such instruments were not particularly efficient, but by the second half of the eighteenth century William Herschel, a Hanoverian-born Englishman, had brought the reflector to a fairly high state of perfection. With one of these instruments, having a mirror with a 6 1/2-inch aperture, he discovered the planet Uranus in 1781, and with still larger instruments he captured two new satellites of Saturn in 1787 and two of Uranus in 1789.

Meanwhile, the refractor had also made a comeback, thanks to the invention of the achromatic lens by Chester More Hall and, independently, by John Dolland, both Englishmen. The idea was to make a composite lens of which one component is crown glass, the other flint. The chromatic aberration produced by the first component is swallowed up by the second so that at least the more serious problems of false color could be eliminated. As a result, a good 2 1/2-inch refractor could be made with a length of only 20 inches, rather than 20 feet as in the days of the old "aerial" telescopes.

The first achromats were limited to about 4 inches in diameter because of imperfections in the quality of the glass then available, but in due time glassmakers improved their techniques, and during the nineteenth century the refractor came once more to dominate the field. For planetary work they are probably still the standard. However, because in the reflector the mirror can be supported from behind (which is impossible with a lens), it is possible to build much larger reflectors than refractors; indeed, the 40-inch refractor of the Yerkes Observatory, near Chicago, though built in 1897, has never been surpassed, and it probably never will be. In the present century, almost all the important instruments have been reflectors.

Other things being equal, the larger the aperture, the finer should be the accessible planetary details. This follows from the optical theory of diffraction, or the interference of light waves passing through an aperture. Even light from a point source such as a star on passing through an aperture forms an image which is a small disk—the larger the aperture, the smaller the disk. An extended object such as a planet may be thought of as consisting of a mosaic of many such points, so the same principles apply. The image formed by the larger telescope should therefore be less "grainy"— and hence more detailed—than that formed by the smaller one.

All this is well and good, but in practice the Earth's atmosphere has as much to do with the quality of the definition as does the aperture. Newton realized this and wrote in the *Opticks* that "If the theory of making telescopes could at length be fully brought into practice, yet there would be certain bounds beyond which telescopes could not perform. For the air through which we look upon the stars, is in a perpetual tremor." Indeed, absolute steadiness of the image is something that one dreams about. Under the usual circumstances, the planet dances in the eyepiece like a confused blob, holding still only now and then. This is a result of undulations in the horizontal layers of the atmosphere. The problems due to these atmospheric disturbances are, in general, more serious the larger the telescope, and a larger telescope will actually produce a blurrier image than a smaller one if the amplitude of the atmospheric undulations is great enough.

Another point worth keeping in mind is that the duration of the "revelation peeps" is generally very brief. This means that the photographic plate, introduced into astronomy in the late nineteenth century and invincible when it came to building up images of faint stars and nebulas, registered only a blur when it came to the most fleeting details on planetary surfaces. So the eye, which is suited by its agility to following the swift undulations of the image, remained supreme in planetary research long after it had been rendered obsolete in other branches of astronomy. Only with the recent advent of charged couple devices (CCDs), which are able to record images with exposures of only 0.05 seconds, has the eye's long reign come to an end.

In many ways, the three and a half centuries after the time of Galileo constituted the romantic period of planetary astronomy. The astronomers who peered at the planets through their long telescopes partook of that piquancy common to all beholders of promised lands that they themselves can never reach, a piquancy that one of them, Percival Lowell, captured perfectly when he wrote of himself observing Mars with his telescope in Arizona in 1894:

> Against hope hoping that mankind may
> In time invent some possible way
> To that far bourne that while I gaze
> Seems in its shimmer to nod my way.

Valiant as they were, the telescopic observers struggled against well-nigh impossible odds. For definitive views of the planets it is necessary, of course, to rely on spacecraft—the "possible way" that Lowell hoped would one day exist.

CHAPTER THREE

Rockets into Space

Ever since it was realized that the planets were other worlds, human beings have dreamed of traveling to them. As early as the second century A.D., a Greek writer, Lucian of Samosata, described a voyage to the Moon by a ship launched by a waterspout—the impracticality of the method rendering unnecessary his apologetic preface: "I write of things which I have neither seen nor suffered nor learned from another, things which are not and never could have been, and therefore my readers should by no means believe them." The great astronomer Kepler, in his posthumously published *Somnium* (Sleep) of 1635, told of being carried to the Moon by demons, while three years later the English bishop Francis Godwin described a lunar voyage in which his Spanish hero, Domingo Gonzales, arrived on the Moon by means of a raft pulled by wild swans.

Rather more plausible is the idea found in Jules Verne's 1865 tale *From the Earth to the Moon*. Verne's travelers are carried to the Moon in a projectile fired from a giant "space gun" (by a remarkable coincidence, Verne located this in Florida not far from Cape Canaveral). Verne made a point of sticking to the facts wherever possible; thus his projectile was fired at a speed of 11 kilometers per second, the escape velocity of the Earth, that is, the speed at which a projectile must travel if it is to escape into space instead of falling back to the Earth.

At least Verne had a kernel of the right idea, and as we shall see, it inspired others to take the next step. But in practice his method was as farfetched as Lucian's waterspout or Godwin's swans. No one could withstand the violence of being fired from such a gun, and in any case, the projectile would burn up in the atmosphere during the ascent.

The real "father of space-travel" was not Verne but a Russian, Konstantin Tsiolkovsky. It was he who first proposed a practical method of traveling outside the Earth's atmosphere: the rocket. Tsiolkovsky was born at Izhevsk in Russia in 1857. When he was nine he came down with scarlet fever, which left him permanently deaf and nearly killed him. He had to leave school and from then on was completely self-taught. In his early twenties he entered upon a career as a provincial schoolteacher, moving first to Borovsk and in 1892 to Kaluga, where he became an instructor in physics at the Ladies Parish Local School—hardly a leading scientific center.

In Tsiolkovsky's time, rockets were hardly a novel invention. They had already been around for many centuries. They were invented by the Chinese, who filled them with another of their inventions, gunpowder, and used them as weapons against the Mongols as early as the thirteenth century. An Englishman, William Congreve, developed them for use during the siege of Copenhagen in 1807, and they also figured in the War of 1812.

"For a long time," Tsiolkovsky wrote, "I thought of the rocket as everybody else did—just as a means of diversion and of petty everyday uses. I do not remember exactly what prompted me to make calculations of its motions. Probably the first seeds of the idea were sown by that great fantastic author Jules Verne—he directed my thought along certain channels, then came a desire, and after that, the work of the mind." Tsiolkovsky realized that a rocket being propelled forward is simply a physical expression of Newton's law of action and reaction: Any action produces an equal and opposite reaction. As an illustration, consider what happens when the air is let out of a balloon. The gas escapes through the nozzle; as it rushes backward, the balloon is pushed ahead. In the same way, combustion of the fuel of a rocket—gunpowder, for example—produces hot gases that rush out of the exhaust at high speed, propelling the rocket forward. The important point is this: rocket propulsion is in no way dependent on the presence of the Earth's atmosphere. Provided one carries along a supply of oxygen and fuel—the necessary ingredients for combustion—there is nothing to prevent a rocket from working perfectly well in a complete vacuum. Indeed, it will work even better there than within the Earth's atmosphere, since the rocket's forward movement would not be impeded by air resistance.

Tsiolkovsky started thinking along these lines in the 1880s, though his first publication on spaceflight did not appear until 1902. In this and subsequent papers he showed the advantages of using liquid fuels like kerosene instead of gunpowder, and he calculated that the most efficient fuel of all would be hydrogen. He was also the first to point out the greater efficiency that could be achieved by building multistage rockets, dreamed of solar power, and designed visionary space stations in which the effects of gravity were simulated by rotating the station. In every respect he was far ahead of his time, and a few years before his death in 1934 he was made a national hero. A meeting of the Soviet Academy of Sciences was held in his honor, and Stalin himself sent him a congratulatory telegram.

The next step was taken by an American, Robert Goddard, who was born in Worcester, Massachusetts, in 1882. Goddard worked in complete ignorance of what Tsiolkovsky had been doing—in the West, Russia was regarded as a backward country, and no one paid any attention to what was going on there. Goddard went to Clark University, in Worcester, where, after taking his doctorate, he became a professor of physics. Right before World War I he became interested in rockets and worked out the main principles of rocket flight. Some of his results were published in 1919 in the treatise *A Method of Reaching Extreme Altitudes*. Goddard was mainly concerned with the idea of using rockets for upper atmospheric research, but he also had a far more startling idea: that a rocket might actually succeed in reaching the Moon and that such a vehicle, if loaded with flash powder, might be seen from Earth when it crashed on impact. Predictably enough, Goddard was viewed as a crank, and in an editorial in the *New York Times* the following year, he was even accused of ignorance of the most elementary principles of physics in imagining that a rocket could work in a vacuum.

Fortunately, Goddard was not discouraged and went ahead with his work, though now with a determination to avoid publicity. Tsiolkovsky had been strictly a theorist, but Goddard actually built the rockets he dreamed of. In his 1919 treatise he had concentrated on solid-fuel rockets, but in the following years he turned to liquid fuels, and on March 16, 1926, on the farm of his Aunt Effie near Auburn, Massachusetts, he successfully fired the first liquid-propellant rocket. Its flight lasted 12 1/2 seconds. Later Goddard built more powerful rockets, and in the 1930s he moved his base to a ranch near Roswell, New Mexico. In the meantime, others had entered the field. In Germany, Hermann Oberth, a Romanian-born schoolteacher working independently of Tsiolkovsky and Goddard, published *Rakete zu den Planetenräumen* (The rocket to interplanetary space) in 1923. Through its influence the Verein für Raumschiffahrt, or

Society for Space Travel, was organized in Germany, of which two of the leading members were Wernher von Braun and Willy Ley.

The members of the society began firing their own liquid-propellant rockets from a Berlin field in 1930, knowing nothing of Goddard's earlier experiments. But while Goddard worked alone, the German army soon grasped the rocket's military potential, and in 1932 von Braun was called to Kummersdorf to become a consultant to the army's new rocket station there. After the Nazis came to power, von Braun was placed in charge of a much larger complex, at Peenemünde in the Baltic, where the deadly V-2 rocket bombs were developed. Though they had only one stage, they were capable of reaching an altitude of over 150 kilometers, and during the last year of the war they inflicted terrible damage on London.

At the end of the war, von Braun and the other German rocket engineers from Peenemünde surrendered to the Americans, and an American rocket program was quickly organized under their leadership. Several captured V-2s were fired from White Sands, New Mexico, and improved rocket designs were also developed—the first multistage rocket, for example, which consisted of a V-2 rocket for the first stage and a smaller rocket known as the WAC Corporal for the second stage, was successfully fired in 1949, reaching an altitude of 400 kilometers.

The V-2 had clearly demonstrated the military potential of the rocket, and this of course remained the chief interest, but in July 1955, the United States announced plans to launch the first artificial satellite, or manmade "moon," called Vanguard, into orbit around the Earth. At the time, American technological superiority was taken for granted, and it was generally expected that the United States had the field to itself. But the Russians also had a keen interest in rocketry that dated back to Tsiolkovsky. Moreover, despite the fact that by the time they arrived at Peenemünde at the end of the war everything of value had been captured by the Americans, and the facility itself was destroyed, they were able to design their own V-2 type of rocket models. The leading role was taken by a brilliant engineer, Sergei P. Korolev, whose own experiments with rockets had begun in the 1930s. In short, the Russians were farther ahead than anyone in the West realized, and on October 4, 1957, Korolev and his team of engineers launched Sputnik 1, the first artificial satellite, from the Russian launch facility at Baikonur, in central Kazakhstan.

The idea of an artificial satellite is straightforward enough. Consider what happens to a body that is thrown upward from the surface of the Earth, whether by a rocket or by any other means. In the absence of any force acting on it, it would travel in a straight line. In fact, however, air resistance and the gravitational pull of the Earth bend its path into a curve. If the velocity is low, the body will simply follow an arc back to the Earth. At a higher velocity, the curvature of the arc becomes gentler, and at a certain point—when the velocity is 28,000 kilometers per hour (km/hr), to be exact—the body will be traveling forward and dropping downward at just the rate that the Earth's surface below curves away from it. Thus, although the body is falling freely, it never reaches the ground (for the moment, we are ignoring air friction). This is why astronauts experience weightlessness in orbit; it is not because they have completely escaped from the Earth's gravity but rather because they are then in a state of free fall.

Sputnik 1 was a smallish sphere, weighing less than 100 kilograms, and it was far too small to be visible with the naked eye; those who seem to recall viewing it as it coursed across the sky were actually seeing its much brighter "booster" stage. There was, however, a radio transmitter on board, and the beeping signal it broadcast was heard around the world. The orbit of Sputnik 1 was an ellipse. At its farthest point from the Earth, or apogee, it was nearly 1,000 kilometers out; at its closest point, perigee, it approached to within 240 kilometers, where there is still an appreciable density of air. At this altitude there is drag on the satellite due to atmospheric friction, and over time its orbit decays. Indeed, after only three months in space, Sputnik 1 fell back to Earth, burning up like a meteor in the atmosphere on January 4, 1958.

Meanwhile, on November 3, 1957, the Russians had launched a much larger satellite, Sputnik 2, which in addition to scientific instruments carried a dog into space. The American rocket team was stunned, and matters became even worse when, on December 6, 1957, the first Vanguard blew up on the launchpad. Finally, on January 31, 1958, the Americans succeeded in putting a satellite into orbit. This was Explorer 1, and though much smaller than either of the Sputniks, it did send back results of greater scientific value, because from its data James Van Allen first identified the radiation belts that surround the Earth and that now bear his name.

The next challenge was to send a spacecraft to the Moon. In principle this is no different from putting a satellite into orbit around the Earth except that it requires a rocket with more thrust. Remember, the Moon is itself under the Earth's gravitational control; sending a rocket there means putting the rocket into a highly elliptical path in which the apogee lies in the vicinity of the Moon. To do so, the rocket must be accelerated to a speed of about 39,000 km/hr. At a distance of 63,000 kilometers from the

Moon, the spacecraft begins to fall within the sphere of influence of the Moon's gravitational pull.

The first attempt to reach the Moon by means of a rocket was the American spacecraft Pioneer 1, launched on October 11, 1958. The satellite reached a height of 115,000 kilometers but then fell back to Earth, burning up in the atmosphere. Later in the same year, Pioneer 2 met much the same fate. Once again it was the Russians who achieved the first success. On January 4, 1959, their Lunik 1 spacecraft passed within only 7,500 kilometers of the Moon, and instruments on board showed that the Moon has no magnetic field. The first spacecraft actually to reach the surface of the Moon was Lunik 2, which crash-landed not far from the crater Archimedes on September 12, 1959—thereby accomplishing what Robert Goddard had been ridiculed for proposing only four decades earlier. Less than a month later, on October 4, 1959, Lunik 3 was launched. It returned the first photographs of the Moon's hitherto unknown face. The photographs were very poor by present standards, but a few features could be clearly identified, of which the most prominent crater was named, fittingly enough, Tsiolkovsky.

So began the great Race to the Moon, which was officially announced in President John F. Kennedy's speech of May 1961 and which was played out against the backdrop of the cold war. Sputnik 1 had been launched when the McCarthy "witch hunts" were still fresh in people's memories, and the first men in space (beginning with Yuri Gagarin's one-orbit flight in Vostok 1 on April 12, 1961, followed by Alan Shepard's suborbital flight in "Freedom 7" three weeks later) shared headlines with the building of the Berlin Wall and the Cuban missile crisis.

The Russians had most of the early "firsts," but they too had their disappointments. After changing the name of their Lunik series to Luna, they suffered five consecutive failures. Meanwhile the American Ranger program, which was intended to provide close-up photographs of the lunar surface during the last few minutes before the spacecraft crashed, was equally hapless. Several spacecraft missed the moon completely, and though Rangers 4 and 6 reached the surface, no useful results were received from them. Finally, in July 1964, Ranger 7 succeeded in sending back close-up photographs from Mare Nubium. Rangers 8 and 9 were also successful, sending back photographs from Mare Tranquillitatis and the large crater Alphonsus early in the following year.

Making a soft landing is, needless to say, a far more difficult feat technically than merely crashing on the surface, and it proved to be yet another Russian triumph. On January 31, 1966, Luna 9 safely touched down in Oceanus Procellarum. The photographs it sent back showed for the first time what the view was like from the lunar surface. The Russians followed this success with Luna 13 in December of the same year. By then the American Surveyor program had gotten underway. Of seven spacecraft launched, numbers 1, 3, 5, 6, and 7 were successful, accomplishing soft landings at various sites between June 1966 and January 1968.

Probes were also put into orbit around the Moon. Though once again the first spacecraft to arrive was Russian—Luna 10 in March 1966—far more important were the five American Lunar Orbiters, which went into lunar orbit between August 1966 and August 1967. They photographed the Moon in great detail, including the far side; in all only a small area near the lunar south pole was not covered. The features shown were much smaller than those visible to terrestrial observers, even those having access to large apertures, and the resulting maps rendered obsolete all previous efforts. Some of the frames were spectacular indeed—Orbiter 1's view of the Earth rising above the lunar wastelands, for instance, and the rugged highlands of the crater Copernicus as seen from Orbiter 2.

During all this time the Russian and American manned programs had been continuing apace, but then disaster struck. In January 1967 three astronauts—Virgil Grissom, Edward White, and Roger Chaffee—were killed in a launchpad fire during testing of the Apollo spacecraft. The Russians also had their setbacks: their chief designer, Sergei Korolev, had died suddenly early in 1966, and the first manned flight under his successor, Vashili Mishin, ended tragically only three months after the Apollo fire when cosmonaut Vladimir Komarov died during the maiden flight of the new Soyuz spacecraft. His capsule crashed at high speed when the lines of his parachute became tangled during his descent.

The Americans had from the outset been quite definite about their plans to land men on the Moon by the end of the decade, and they outlined in detail the steps they intended to take. The Russians were always more secretive; they had taken the early lead, but no one knew just what they were planning to do next.

The next feat planned by both the Russians and the Americans was to put a man in orbit around the Moon. The Russians launched Zond 5 in September 1968. This was an unmanned craft whose mission was to circle the Moon and return to the Earth as a dress rehearsal for a manned mission. Much to the disappointment of Russian engineers, however, the craft came down far off course, in the Indian Ocean. Another spacecraft,

Zond 6, was launched on a similar round-the-Moon mission in November 1968, but again there were no cosmonauts aboard. This was to be the last Russian moon flight until July 1969.

Meanwhile, the Apollo spacecraft, after a nearly two-year delay following the launchpad fire, was again pronounced fit to fly. Apollo 7 made a successful Earth-orbital flight in October 1968, thus setting the stage for Apollo 8, which made the first manned circumlunar flight in December 1968.

For the first time the launch vehicle was the powerful three-stage Saturn V rocket. The first stage used kerosene as its fuel, and the second and third used liquid hydrogen, which had to be stored in specially insulated tanks in order to be kept below the temperature at which liquid hydrogen boils, −253°C. The whole assembly was taller than the Statue of Liberty—though, needless to say, only the relatively small capsule made the full journey to the Moon and back to the Earth again. The Apollo 8 spacecraft, carrying astronauts Frank Borman, James Lovell, and William Anders, arrived in lunar orbit on Christmas Eve 1968, made ten circuits at a height of less than 110 kilometers, and then returned to the Earth.

Apollo 9 was an Earth-orbital flight that involved testing the lunar module, the craft designed for ferrying astronauts to the lunar surface. Apollo 10, which prior to the launchpad fire had been scheduled for the first lunar landing, tested the lunar module in the vicinity of the Moon in May 1969; astronauts Thomas Stafford and Eugene Cernan piloted the craft to within only 15 kilometers of the lunar surface, while John Young remained above in the command module.

This was the last of the planned preliminaries. The next step was to be the actual landing of men on the Moon by the crew members of Apollo 11 in July 1969. The flight to the Moon and entry into lunar orbit were uneventful, but the suspense was increased by the fact that two days earlier an automatic probe, Luna 15, had also arrived in lunar orbit in an apparent last-ditch effort by the Russians to upstage the Americans.

The Apollo crew went ahead with its maneuvers. Separating from Michael Collins in the command module, Neil Armstrong and Edwin Aldrin swept down in the lunar module "Eagle" toward what had appeared to be a smooth landing site in Mare Tranquillitatis. The spacecraft was on automatic controls, but in the last few seconds Armstrong had to take over manually in order to avoid coming down into an unexpected field of boulders. After this brush with disaster came the thrilling words, "The Eagle has landed."

Five hours after touchdown, Armstrong and Aldrin began their "extravehicular activity" on the lunar surface. In all they spent 2 hours, 47 minutes outside the lunar module, during which time they collected 22 kilograms of lunar rocks and deployed various scientific instruments.

Meanwhile, Luna 15 attempted its own soft landing. It had been designed to bring back samples of Moon rock to the Earth but failed completely, crashing into Mare Crisium while the astronauts were still at "Tranquillity Base," and no useful results were obtained from it. The Race to the Moon was over.

Other manned flights to the Moon followed, beginning with Apollo 12 in November 1969. This time Charles Conrad and Alan Bean landed in Oceanus Procellarum very close to the site where Surveyor 3 had been resting since April 1967. The astronauts made the short walk to the probe and removed parts of it for return to the Earth.

The flight of Apollo 13 in April 1970 nearly ended in disaster when an oxygen tank on board the spacecraft's service module exploded on the way out to the Moon. The three astronauts, James Lovell, Jack Swigert, and Fred Haise, could not simply reverse course and return to the Earth. They were committed to circling once around the back side of the Moon, making most of the journey in the lunar module in order to conserve supplies. Only just before reentry did they return to the command module for the descent through the Earth's atmosphere. Fortunately, no one had been killed, and although the explosion clearly had been a major setback, the problem with the oxygen tank was quickly solved so that the launch schedule was delayed by only a few months.

The next flight, Apollo 14, took place in January 1971. Alan Shepard, who had been the first American in space in May 1961, landed with Edgar Mitchell near the hilly region of Fra Mauro. In July 1971 the crew of Apollo 15 explored another rugged area of the Moon, Hadley Rille, near Mount Hadley in the Apennine range; it was the first mission in which the electrically powered lunar rover was used. Apollo 16 landed near the crater Descartes in April 1972, and Apollo 17, the last of the Apollo missions, explored the Taurus-Littrow valley in December 1972. All told, twenty-seven men have been to the Moon, of whom twelve have actually set foot there.

Meanwhile, the Russians were far from idle. They followed the unsuccessful Luna 15 with a series of triumphs. Luna 16 landed in Mare Fecunditatis in September 1970, collecting Moon rocks and bringing them back to the Earth. This was the first time an automatic craft had ever done so. Later in the year came Luna 17. After a soft landing in Mare Imbrium, it deployed Lunokhod 1, a buggylike vehicle capable of roaming the surface and sending back photographs from various points. Several other Lunas followed, of which the most notable were Luna 20 of February 1972, which landed in Mare Fecunditatis 120 kilometers from the Luna 16 site and also returned samples; Luna 21 of January 1973, which carried the second Lunokhod vehicle, this one to Le Monnier crater in Mare Serenitatis; and Luna 24 of August 1976, which returned samples from Mare Crisium.

Of course, the Russians never did attempt a manned lunar landing, and for many years they claimed that they had never intended to do so. Now, however, with the more open mood of *glasnost,* the full story has been told. A manned lunar spacecraft was indeed ready by 1968, but the powerful N-1 booster rocket that was supposed to carry it to the Moon proved to be fatally flawed. Several test launches ended in massive explosions, and in one case a launchpad was completely destroyed. After the Apollos achieved their triumphs, there no longer seemed to be any point in carrying on with the project, and it was unceremoniously scrapped. The engineers consoled themselves with their brilliantly successful unmanned probes.

There is no denying that the Apollo flights rank among humanity's greatest adventures, but American interest was astonishingly shortlived. Nothing was done to follow up on the initial successes, and even the lunar landings were cut short—of the twenty-one missions originally planned, the last four were canceled. Since December 1972 manned spaceflight has been cautiously tethered to the gravitational apron strings of the Earth.

Even before the last Apollo flew, the Russians had launched the first of their Salyut orbital stations, and new stations followed at intervals of about one every other year, for a total of seven in all. The stations were staffed by crews ferried to them by the standard Soyuz craft. A more ambitious space station, Mir, followed in February 1986, and since then cosmonauts have maintained an almost continuous presence on board, with individual cosmonauts remaining aloft for as much as a year at a time. The Americans also deployed a manned orbiting station, Skylab, which remained in orbit from May 1973 until July 1979, when it plunged to destruction in the Earth's atmosphere, though by then, of course, it had long since been abandoned, and there was no one on board.

The next development was the American reusable manned spacecraft, or Space Shuttle, of which the first launch took place in April 1981. When fully deployed, three Shuttles were to deliver payloads into space on a routine basis. For several years all went well, but then came the Challenger explosion in January 1986, which killed seven astronauts. Inevitably a distinct note of caution set in. A number of important launches had to be pushed back while the causes of the disaster were investigated and corrective measures taken to redesign the faulty components and faulty decision-making procedures. The Space Shuttle is now flying again, but the program has continued to have problems, and in retrospect there is little doubt that American space planners made a disastrous mistake in choosing to rely exclusively on manned spacecraft for launching payloads into space. The main purpose of the Challenger flight, it should be recalled, was simply to put a routine communications satellite into orbit, something that had been done countless times before with expendable vehicles—and with no risk to human life.

Plans are now moving forward to build a large-scale orbiting space station, and there have been stirrings about setting up permanent bases on the Moon and making a manned flight to Mars. There are, however, critics who suggest that it would make far more sense to put the emphasis solidly on unmanned probes. Consider, for example, what would be involved in the mission to Mars. The straight-line distance to Mars is never less than 140 times as great as that to the Moon, which means that a one-way journey would take many months. Thus a program to send people to Mars would inevitably be far more difficult and expensive than the Apollo program, and it has been argued—correctly, I think—that if the main goals are scientific, these could be accomplished much more efficiently by means of remote-controlled probes similar to the Russian Luna series of the 1970s.

Whether or not a manned expedition to Mars is mounted within the next several decades, it is clear that, on the whole, humans are not really very well suited to withstanding long exposure to the hostile environments of space. Romantic notions aside, people are an encumbrance. The conquest of deeper interplanetary space will belong, for the foreseeable future, to our machines.

In order to reach the planets, a spacecraft, whether manned or unmanned, must first break free of the Earth's gravitational pull, and to do this it must be carried to a speed of 11 km/sec, or 40,000 km/hr. At that point it has become an independent body traveling in its own orbit around the Sun. The path it follows to another planet is called the transfer orbit. In order to travel to a planet lying inward toward the Sun—Venus, for example— it is necessary to slow the spacecraft down relative to the Earth. To travel to a planet farther out, like Mars, the spacecraft must be sped up.

There have now been many interplanetary probes, going back to Venera 1, which the Russians launched in February 1961. Its destination was Venus, but radio contact was lost long before it reached the planet. The Russians also made the first attempt to reach Mars with their Mars 1 spacecraft, launched in November 1962, but that also ended in failure. In both cases it was the Americans who achieved the early successes, with Mariner 2 which passed by Venus in December 1962 and Mariner 4 which encountered Mars in July 1965.

There have been many dramatic missions since. The Russian Veneras, for example, made successful landings on Venus in the early 1970s. The American Mariner 9 went into orbit around Mars in 1971, followed by the two Vikings, which made the first soft landings on the surface in 1976. In 1974, Mariner 10 became the first, and so far the only, spacecraft to visit the innermost planet, Mercury.

The giant planets were first reached by the two Pioneer spacecraft of the 1970s. This set the stage for the two Voyager spacecraft, launched in 1977. Because of a rare planetary alignment—it will not be repeated until 2152—the Voyagers were able to be slingshotted gravitationally from one planet to the next, which not only shortened by many years the time needed for the trip but also allowed a single spacecraft to survey several planets. Both Voyager spacecraft passed by Jupiter and Saturn, while Voyager 2 alone went on to Uranus and finally Neptune, which it reached in August 1989 after a twelve-year odyssey.

Few will need to be reminded of the breathtaking images sent back from the spacecraft, showing the riotous clouds of Jupiter, the dramatic volcanoes of Io, the complex rings of Saturn, the icy canyons and cliffs of Miranda, Neptune's deep-blue oceans of methane, and the cantaloupe-rind surface of Triton.

The 1980s saw a dearth of American interplanetary initiatives—remember that the Voyagers, though achieving some of their greatest triumphs during the 1980s, were launched as long ago as 1977. Fortunately, the 1990s opened on a more promising note. The Magellan spacecraft, designed to carry out detailed radar mapping of the surface of Venus, entered orbit around that planet in August 1990, and after some initial problems in maintaining contact, worked splendidly. Meanwhile, the Galileo spacecraft is on its way to Jupiter via a threefold gravitational assist from Venus and the Earth. The spacecraft bypassed Venus on February 10, 1990, made a first flyby of the Earth on December 9, 1990, and is scheduled to make a second on December 8, 1992, which will propel it on to Jupiter. Though interesting photographs of the far side of the moon were sent back using a low-gain antenna during the December 1990 flyby, the spacecraft's main antenna—on which it will depend to complete its objective at Jupiter—has so far failed to deploy. As of December 1991 it remained jammed.

Both Magellan and Galileo were successfully launched from Space Shuttles. So was the Hubble Space Telescope, which, despite having an incorrectly figured primary mirror, has returned some breathtakingly detailed images of the planets.

Other countries have also been active. No fewer than five spacecraft, for instance (two Russian, two Japanese, and one European), rushed out to meet Halley's Comet in 1986, and the Russian Phobos mission sent back interesting results from Mars in March 1989 before radio contact was suddenly lost. (However, recent political events—namely, the disintegration of the Soviet Union—make it unlikely that the Russian space program will be able to maintain its leadership role of former years.)

Other spacecraft will no doubt follow; indeed, ambitious missions to Mars and Saturn are already planned for the coming years. It is probably fair to say, however, that the first survey of the Solar System has been completed. We can be quite sure now, as we could not be as recently as 1960, that we have the main outlines right. Let us, then, look at the Moon and planets one by one to see what has been discovered about each of them.

CHAPTER FOUR

The Moon

The Moon, at a distance of only 385,000 kilometers, is the celestial body of which the features of the surface can be scrutinized in greatest detail from the Earth, and so far it is the only one on which human beings have actually set foot. Even with the naked eye, the Moon's large dark patches are readily visible; they make up the face of the "Man in the Moon." Leonardo da Vinci, who died in 1519, long before the invention of the telescope, attempted to sketch these markings, and in his notebooks he hinted at a theme that would prove to be a singularly enduring one—that of lunar change:

> If you keep the details of the spots of the moon under observation you will often find great variation in them, and this I myself have proved by drawing them. And this is caused by the clouds that rise from the waters in the moon, which come between the sun and those waters, and by their shadow deprive these waters of the sun's rays. Thus those waters remain dark, not being able to reflect the solar body.

The first man to turn a telescope toward the Moon was not Galileo, as is commonly believed, but England's Thomas Harriot. He made a sketch of the Moon in July 1609—several months before Galileo—though it shows very little detail. Galileo's drawings were far better, and moreover he conducted his researches with such thoroughness that he certainly deserves the main credit. In addition to his drawings of the lunar formations, Galileo established that the lunar surface is very rough, and he even measured the heights of the lunar mountains—his estimate of four Italian miles (about 6,000 meters) was of the right order of magnitude.

The first maps of the Moon began to appear during the generation after Galileo. Hevelius at Danzig produced a tolerably good chart in 1647 on which he named the large dark areas after terrestrial seas (thus there came to be a Mediterranean and an Adriatic on the Moon), while the craters received the names of mountains (Mt. Etna and Mt. Sinai, for example). The crater that later came to be called Plato was an exception, being named, because of its dark floor, the Greater Black Lake.

Another map was drawn up in 1651, by the Jesuit astronomer Giambattista Riccioli on the basis of observations made by a junior colleague of his, Francesco Grimaldi. This map is important because on it Riccioli introduced a novel scheme of nomenclature that eventually became official and that is the basis of the system still in use today.

It is worth considering some of Riccioli's views about lunar conditions. As recently as 1633, Galileo had been forced to recant before the Roman Inquisition, and thereafter it had become unlawful for Catholics to profess the Copernican theory. Riccioli, as a Jesuit, was bitterly opposed to the idea that the Earth traveled around the Sun and instead introduced a variant on the Tychonic theory in which Mercury, Venus, and Mars moved about the Sun, while the Sun, Jupiter, and Saturn moved about the Earth. Thus Riccioli held that the Earth was at rest in the center of the universe. Instead of being an ordinary planet, as Copernicus had maintained, it was unique—and so, presumably, was man. This was Riccioli's firm view, and he denied that there could be either water or inhabitants on the Moon. On his lunar map he deferred to tradition, however, naming the dark areas maria, or seas. Perhaps he was indicating his reservations in the fanciful names he chose—Mare Imbrium (Sea of Rains), for instance, and Oceanus Procellarum (Ocean of Storms), Mare Tranquillitatis (Sea of Calm), Mare Serenitatis (Sea of Peace), Lacus Somniorum (Lake of Dreams), and Sinus Iridum (Bay of Rainbows).

Riccioli gave the lunar craters the names of philosophers, scientists, and mathematicians, with his selections betraying his prejudices. The most prominent crater honored Tycho, and Ptolemy also received a splendid feature, Ptolemaeus, while the proponents of the heliocentric theory—Aristarchus, Copernicus, and Kepler—were tossed into the Ocean of Storms. Riccioli honored several of his fellow Jesuits with large formations, including Fathers Clavius, Scheiner, and Grimaldi, and he reserved a choice feature for himself near the moon's western limb (the western edge of the visible disk) beside another that he named for Grimaldi. For the most part, his choices were wisely made, and because of the convenience of his system—or the vainglory of astronomers—it eventually superseded the awkward system of Hevelius.

About 1750 the German astronomer Tobias Mayer drew up an excellent small chart of the Moon. Mayer also planned a larger chart, but it was left unfinished at his untimely death in 1762 at the age of only thirty-nine, and even the small chart did not appear until 1775. Mayer was also the first to give a complete explanation of the so-called librations of the Moon, a phenomenon first noted by Galileo. This is an important concept, and it deserves a brief digression.

Though the Moon's rotation with respect to the Earth is captured, the Earth-based observer actually sees the Moon wobble slightly relative to the Earth-Moon line. Two different wobbles can be distinguished. An east-west wobble, or libration in longitude, is due to the fact that though the Moon maintains a constant angular rate of spin, its speed in its orbit varies, as described by Kepler's second law. Thus the motions of rotation and revolution get slightly out of step with each other, and the Moon rocks gently back and forth. At the same time, there is a north-south wobble, or libration in latitude, which is due to the fact that the Moon's orbit is inclined slightly with respect to the Earth (the inclination is actually somewhat variable, ranging from 4° 58′ and 5° 19′, but this does nothing to change the basic argument). As a result of these librations, an observer is sometimes allowed to peek over to regions of the lunar far side, and somewhat more than half of the lunar surface—59 percent to be exact—can be scanned from the Earth, though admittedly features at the extreme limb are always poorly visualized because of optical foreshortening. Until the Lunik 3 probe flew around to the back side of the Moon in 1959, the other 41 percent of the lunar surface was completely unknown.

The Moon's rotation was much more rapid in the remote past. Gradually the tidal force of the Earth on the Moon slowed it down to its present rate. Similarly, the Moon exerts a tidal force on the Earth which is slowing its rotation and lengthening our day. Some 300 million years ago there were 400 days in a year, and in the far future the Earth's rotation will be locked with the Moon (as Pluto's now is with Charon), taking fifty days to rotate once on its axis as the Moon completes its orbit in the same period.

With Mayer, the first reconnaissance of the Moon was complete, but a much more detailed study commenced with the work of Johann Hieronymus Schroeter. Born in Erfurt in 1745, he studied law at the University of

Göttingen and became acquainted with the Herschel family in Hanover, where he served as a government official. Deciding that astronomy was his real passion, he obtained a position as chief magistrate at Lilienthal, a sleepy little village near Bremen where he would have time to devote to his astronomical work. Indeed, he established the foremost observatory on the Continent there. His largest telescope was a reflector with a 27-foot focal length and a mirror 20 inches in diameter, but his best work was done with a 13-foot reflector having a mirror of 9 1/2 inches.

The Moon was always Schroeter's special interest, and he proved to be an indefatigable observer. Admittedly, he was a clumsy draughtsman, and some of his ideas sound strange today, but there is no denying that his work represented a significant advance. For example, he was the first to give serious attention to the rilles, which appear like cracks on the surface. Whereas Christiaan Huygens had recorded the Hyginus rille and one or two others in 1685, Schroeter added about a dozen more. However, he believed not only that the Moon had an atmosphere but also that changes took place on the surface—on one occasion, when recording a crater that he had not seen on a previous night, he hastily concluded that it had just erupted onto the surface (see fig. 4.1), while he also believed strongly that there were tracts of vegetation there. Sadly, his observatory was overrun by the French in 1813, during the Napoleonic Wars, and he died three years later.

Others, however, carried on where Schroeter left off. One was Franz von Paula Gruithuisen, a keensighted astronomer who became director of the Munich Observatory in 1826. Like Schroeter, he made a close study of the lunar rilles, concluding that they were dry riverbeds or roads, and he also agreed with Schroeter as to vegetation on the Moon and the presence of a substantial atmosphere. He even went so far as to announce the discovery of a lunar city at the southern edge of Sinus Aestuum—"Oh, Schroeter," he exclaimed, "here is what you have always sought in vain"—but though the region is one of peculiar formation, it does not really look in the least bit artificial.

A soberer study of the Moon was made by a contemporary of Gruithuisen, Wilhelm Gotthelf Lohrmann, a Dresden surveyor. He took a skeptical view of Gruithuisen's alleged discoveries, and instead of seeking evidence of lunar inhabitants, devoted himself to carefully mapping the lunar surface. He planned a chart on a scale of 97.5 centimeters to the Moon's diameter and actually published the first four of twenty-five sections—they were exquisite—but then, for some reason, he gave up. But his fragment served as the inspiration for the work of Wilhelm Beer and

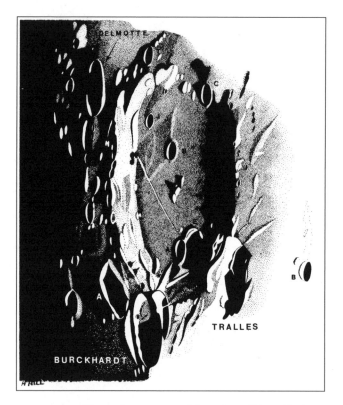

FIGURE 4.1. Cleomedes at sunset. Note the striking rille in
the floor of the crater. This crater was the subject of a long
and careful study by J. H. Schroeter, who recorded changes in
the floor that he considered evidence of volcanic activity. We
now know that his conclusion was incorrect. (Pen and ink
drawing by Harold Hill, based on observations made on
Nov. 4 and 5, 1990, with an 8 ¼-inch Schmidt-Cassegrain
and a 10-inch Newtonian reflector)

Johann Heinrich Mädler, whose study of the Moon was to be nothing less
than revolutionary.

Beer was a Jewish banker who in 1824 began to take lessons in astronomy from Mädler, a schoolteacher in Berlin. Beer was fascinated and soon
acquired a first-rate telescope, a 3 3/4-inch Fraunhofer refractor, which he
set up in a small observatory in the Tiergarten. From 1830 onward the
two men worked diligently at mapping the Moon, with Mädler doing
most of the observing and drawing. Between 1834 and 1836 they issued a
complete map on the same scale as Lohrmann had planned, and they

followed this up in 1837 with their celebrated book *Der Mond,* which presents an essentially modern view of the lunar surface.

In their view, most if not all of the reported changes could be explained as optical effects rather than as the results of vegetative or geological processes. The lunar surface appeared to be bone dry, and as the shadows cast by the lunar features were always jet black and sharply sculpted, they knew that there was little or no air. Thus their conclusion: "The Moon is no copy of the Earth."

This result led, naturally enough, to a relative neglect of the Moon in the years after 1837. The Moon, after all, had been pronounced on the highest authority a dead and useless waste, so there was little reason to observe it on a regular basis. The situation changed after 1866, however, when the German astronomer Julius Schmidt reported that Linné, a small crater in Mare Serenitatis, had a markedly different appearance from what it had in the days of Lohrmann and Beer and Mädler. Apparently, he thought, a volcanic eruption had taken place. We now know that Schmidt was mistaken; spacecraft photographs show Linné to be an ordinary-looking impact crater, and no change can possibly have taken place. But at least the somewhat paralyzing spell that Beer and Mädler had cast over lunar studies was broken, and once again it became respectable to look for—and report—evidence of lunar change.

Occasional veils and red glows on the Moon became—indeed, continue to be—the subject of active discussion, and it may be that the Moon is not altogether dead. But though minor events no doubt sometimes happen—a meteorite fall every now and then, and possibly an occasional outgassing from the crust—major alterations are unknown. Nevertheless, the Moon, with its dramatic relief, is anything but dull. Change there is, and though only of the apparent kind, it is breathtaking. Because of the continuous transformation in the appearance of the rugged surface owing to shadow effects, the observer who overcomes a prejudice about seeing a static face will be delighted to find, to the contrary, that the lunar surface is incredibly protean. Moreover, there is an almost intimidating abundance of detail. As Homer sings in a different context in the *Iliad*:

> I could not tell over the multitude of them nor name them,
> Not if I had ten tongues and ten mouths, not if I had
> A voice never to be broken and a heart of bronze within me.

The worst time to look is at full moon, when the surface appears washed out and suffers badly from a lack of contrast. It is better to study features when they are on the terminator, the line dividing day and night,

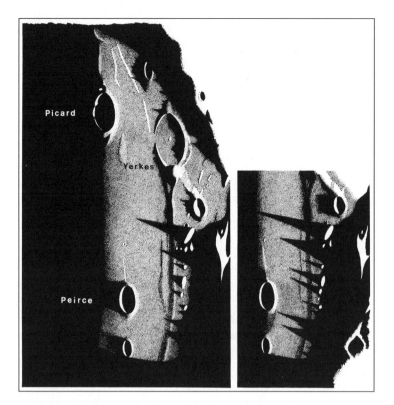

FIGURE 4.2. The west-northwest border of Mare Crisium at sunset, showing lengthening shadows across the mare basin. (Pen and ink drawings by Harold Hill, Nov. 23, 1983, with a 10-inch Newtonian reflector at 190×. The drawing on the left was based on observations made at 00:05–00:45 UT; that on the right, at 01:15–01:35 UT)

rather than under the direct illumination of high noon. Naturally, the most elevated regions are the first to catch the morning sunlight and the last to be swallowed up in the evening darkness. This explains why the terminator looks so jagged in a telescope; irregularities of the surface stand out in sharp relief, their patterns etched in dramatic chiaroscuro and changing continuously with the crawling shadows (fig. 4.2).

After Beer and Mädler, a number of astronomers devoted a good part of their lives to detailed mapping of the lunar surface. Schmidt himself published a large chart in 1878, and valuable work was done in England by such energetic observers as W. R. Birt, T. Gwyn Elger, and Walter Goodacre. Meanwhile, photography was developing into an important tool of astronomical research. Early photographic atlases of the lunar surface were produced at about the turn of the century by M. Loewy and P. Puiseux of France and W. H. Pickering of the United States, and a later atlas was completed by G. P. Kuiper. One of the later visual mappers of the Moon was H. P. Wilkins, an English amateur who drew up a 300-inch chart. He died in 1960 at the dawn of the spacecraft era. Then there was the U.S. Air Force mapping project of the early 1960s, which provided a base chart for the Apollo missions.

All of this work was, needless to say, rendered obsolete by the Lunar Orbiter and Apollo missions of the 1960s, which mapped the surface in exquisite detail. At long last the lunar map was essentially complete, and this brings us to the next question: How did the lunar surface come to look the way it does?

The great craters are, of course, the dominant features of the Moon and are what make its appearance so unlike the Earth's. Galileo was reminded of the "eyes" on a peacock's feathers when he saw them with his small telescope in 1609. A less fanciful comparison might be with volcanic calderas on the Earth, and the idea early arose that similar processes may have been responsible for their formation—that is, that the lunar craters were the result of volcanism.

The earliest version of a volcanic theory was proposed by Robert Hooke of the Royal Society in 1665. On heating alabaster powder to bubbling, Hooke noted that, when the bubbles burst, the resulting structures bore at least a superficial resemblance to the lunar craters. This theory failed to stand the test of time. Even on the lunar surface, where the gravity is only one-sixth that of the Earth, it was impossible to imagine any material that could form a bubble hundreds of kilometers across. Ordinary volcanoes, however, seemed reasonable enough, though there were important differences between lunar craters and terrestrial volcanoes, as the American geologist G. K. Gilbert noted in 1892:

> Ninety nine times out of a hundred the bottom of a lunar crater lies lower than the outer plain; ninety nine times out of a hundred the bottom of a Vesuvian crater lies higher than the outer plain. . . . The lunar crater is sunk in the lunar plain; the Vesuvian is perched on a mountain top. The rim of the Vesuvian crater is not developed, like the lunar, into a complex wreath, but slopes outward and inward from a simple crest-line.

The main rival of the volcanic theory was the meteorite impact theory. Interestingly, Hooke himself had come close to such a theory in 1665. He had noted the similarity between the lunar craters and the structures formed when lead shot was dropped into a mixture of pipe clay and water, but he dismissed the idea because he could not imagine where such bodies could have come from in the case of the Moon. Following the discovery of the asteroids between Mars and Jupiter—the first of which, Ceres, was found in 1801—a possible source was identified, and it is no coincidence that the meteorite theory was first advanced at about that time by Germany's Bieberstein brothers. It later received support from Gruithuisen.

Here too, however, there were objections. Incoming meteorites ought to strike the Moon at all possible angles. By chance, most of them would arrive obliquely, in which case it seems that elongated scars ought to be the norm. But the lunar craters are almost all nearly circular (because of optical foreshortening, craters appear elliptical near the limb, but when this is taken into account, the true forms are found to be circular). Then, too, it was imagined that incoming meteorites would behave like bullets fired into a target. In that case, as E. M. Antoniadi argued in 1897,

> seeing that a shot made a hole its own size in a target, . . . only a bullet 142 miles in diameter would give rise to a depression like Clavius. . . . Meteors therefore of 300, 420, and 700 miles would be required to produce the Maria Crisium, Serenitatis and Imbrium respectively; but . . . our moon would long ago have been "blotted out of the face of the sky" were she to have ever come to grief with such imaginary roving masses.

The debate between the volcanic and meteoritic theories raged on for more than a century. It has now at last been settled: though there are volcanic features on the Moon (see below), the craters as well as the large basins were formed by meteorite impacts.

The true picture began to emerge gradually in the years following World War II, when it became clear that the lunar craters were not analogous to bullet holes after all; instead they closely resemble the craters formed by exploding bombs. In an explosion, the incoming projectile need not be extremely large to form a crater of great size. Indeed, since the energy of such a projectile is given by $1/2mv^2$, the velocity is more important than the mass. A projectile traveling at 30 km/sec, for instance, would need to be only 1 kilometer across to form a crater the size of 97-kilometer-wide Copernicus, one of the most dramatic impact features on the Moon (fig. 4.3). Such an object, when suddenly stopped on hitting the Moon, would transfer its enormous energy almost instantaneously to the surface. Some of the energy would vaporize the meteorite itself and melt rock at the

FIGURE 4.3. The great impact crater Copernicus, 97 kilometers wide, looms on the horizon in this oblique view obtained by the astronauts of Apollo 17 in December 1972. (NASA photograph)

immediate point of impact, but most of the energy would go to producing a shock wave expanding radially from the point of impact. The shock wave excavates the crater and throws out a rim of material around it (the ejecta blanket). The important point is that it is the shock wave that actually forms the crater. This explains how the resulting crater can be circular even in the case of an oblique impact.

Examples of impact craters are not uncommon even on the surface of the Earth. The famous Meteor Crater in Arizona, a little over a kilometer across, is the best known, but in recent years many others have been identified, including a large number on the Canadian Shield (for example, water-filled Manicouagan, 70 kilometers across). But most of the Earth's impact craters have been erased or at least significantly degraded by erosion. On the Moon, which lacks the Earth's wind and water, features from the primitive history of the Solar System are still plainly visible.

Currently, a 1-kilometer crater is formed on the Earth only about once every two or three hundred thousand years and a 100-kilometer crater once every billion years. These rates are far too low to account for the heavily cratered surfaces of the Moon and other planets, which means that space must have been much more cluttered in the early history of the Solar System. That the debris came in all possible sizes is attested by the varying diameters of the craters, which range from microscopic pits to structures hundreds of kilometers across.

The form of the crater also varies according to the size. Small craters, on the order of 10 kilometers or less in diameter, are simply bowl-shaped depressions. The crater in Arizona is of this type. Many of the craters more than 20 kilometers across show a central peak and a terraced rim, and somewhat larger craters show a cluster of central peaks. Between 140 and 300 kilometers, the predominant form is the ringed crater, while the very largest features have multiple rings. (It is not always realized, by the way, how shallow the larger lunar craters are. For example, the crater Copernicus, 97 kilometers wide, has a floor only 3 to 4 kilometers deep, and its walls rise only a kilometer above the surrounding plain, so in profile it resembles a shallow saucer. To someone standing in the middle of such a structure, the walls would be invisible below the horizon.)

The crater-saturated highland regions of the Moon are the most primitive lunar landscapes. Following the accretion of the Sun and planets from the primitive nebula some 4.6 billion years ago, leftover debris continued to roam through interplanetary space. As the planets traveled in their orbits around the Sun, collisions were frequent. The rocks brought back from the lunar highlands by the Apollo astronauts range in age from 4.0 to 4.3 billion years and are of the brecciated variety, made up of fragments jumbled together during this period of incessant bombardment.

Toward the end of this period, some particularly large bodies fell into the Moon to produce the large circular basins. From lunar rock samples it has become possible to assign rather exact dates to some of these features: Mare Nectaris, 3.92 billion years; Mare Serenitatis, 3.87 billion years; Mare Crisium and Mare Imbrium, 3.85 billion years. The Imbrium impact was especially violent. The wreath of mountains, including the Apennine range, whose peaks reach heights of 4,500 meters, defines the basin's outline. They were not produced by folding of the crust like the mountains of the Earth but instead represent the ejecta blanket from this catastrophic event.

Mare Imbrium is an example of a multiringed basin. The Apennine range itself is the outer rim, the inner rim consisting of the more modest

FIGURE 4.4. Mare Orientale, the best-preserved multiringed basin on the Moon. (NASA photograph)

Spitzbergen range. A better-preserved example of a multiringed basin is Mare Orientale (fig. 4.4; this is the "Eastern Sea," so called despite the fact that it lies on the western limb of the Moon; its name is a throwback to the days when lunar compass points were defined by reference to the inverted telescopic image). Though smaller than 1,300-kilometer-wide Mare Imbrium, Mare Orientale shows concentric rings of mountains that are complete and very well formed. Measured across the outermost ring, the diameter is some 900 kilometers. Because from the Earth it is seen only under conditions of extreme libration, Mare Orientale's remarkable structure did not become apparent until the spacecraft era, and the old observers regarded its concentric rings as ordinary mountain ranges, naming the outer ring the Cordillera range and the inner two rings the Rooks. Several other large multiringed basins are found on the lunar far side— notably, Hertzsprung, Korolev, Apollo, and the dark-floored Mare Moscoviense, as well as the remnants of a 2,000-kilometer-wide basin, the largest on the Moon, which extends from the south pole to the crater Aitken and which was first clearly imaged by the Galileo spacecraft in December 1990.

Following the formation of Mare Orientale, the youngest of the great basins at 3.80 billion years, lava that had been flowing out onto the Moon ever since its formation began to accumulate on the near side (earlier basin impacts had broken up or buried earlier flows). The lava filled the deep basins—first Oceanus Procellarum, then Imbrium, Fecunditatis, Nectaris, Tranquillitatis, and Crisium. This did not happen all at once; the basaltic rocks the astronauts brought back from the great basins show a range of ages. One might expect this in any case because of the lack of uniformity in the lava fields, which suggests that there were multiple episodes in which lavas having different viscosity flowed onto the surface. Describing these lava fields in 1895, the English observer T. Gwyn Elger noted that even in a small telescope "one cannot fail to remark . . . many low bubble-shaped swellings with gently rounded outlines, shallow trough-like hollows, or, in the majority of them, long sinuous ridges, either running concentrically with their borders or traversing them from side to side."

A number of the larger craters have been flooded by lava and thus have relatively smooth interiors—examples include Lubiniezky (fig. 4.5), Plato, Gassendi, and on the far side, Tsiolkovsky. Wargentin is flooded almost right up to the level of its walls so that it actually forms a plateau. Archimedes, another flooded crater, lies within the Imbrium basin. This proves that it must have formed sometime between the Imbrium impact and the lava flows. The same is true of the beautiful Sinus Iridum (the Bay of Rainbows), whose impact basin cuts into the Alps range, a section of the Imbrium ejecta blanket. Some craters appear almost totally submerged in the lava fields of the maria so that their walls are seen only in outline. These ruined structures are sometimes called "ghost craters." There are several good examples in Oceanus Procellarum—Flamsteed, for instance.

A typical large crater has, as mentioned earlier, one or more prominent central peaks. In some, such as Alphonsus (fig. 4.6) and Gassendi, there are complex systems of cracks in the floors, thought to have resulted from stresses on the crust produced by magma welling up from below. It is worth noting that both Alphonsus and Gassendi are sites where reddish glows have been reported, and in the interior of Alphonsus there are several dark spots, long mysterious to Earth-based observers but now identified as volcanic cinder cones.

Among the other interesting features on the Moon are the rilles. A number of them are within reach of a small telescope. There is, for instance, the great winding Schroeter's Valley near the crater Aristarchus, and there are several fine specimens in the Mare Vaporum–Sinus Medii region, including Hyginus (actually a crater chain run together) and the complex Triesnecker system (fig. 4.7). Also worthy of mention is Hadley Rille, one of several rilles on the border of the great Imbrium basin south

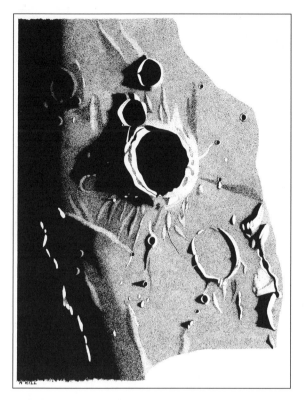

FIGURE 4.5. The Bullialdus and Lubiniezky region in
Mare Nubium. Bullialdus is a typical large impact
crater, measuring some 60 kilometers across.
Lubiniezky, to its lower right, is a good example of a
lava-flooded "ghost" crater. (Pen and ink drawing by
Harold Hill, Oct. 13, 1990, based on observations
made at 4:05–5:30 UT, using an 8 ¼-inch Schmidt-
Cassegrain at 245× and a 10-inch Newtonian reflector
at 286×)

of the crater Autolycus. It is famous for being near the landing site of the
Apollo 15 spacecraft in July 1971. Astronauts David Scott and James
Irwin were able to drive their Lunar Rover almost to its edge.

Some of the rilles follow winding courses, and one can easily understand how Gruithuisen could have imagined them to be dry riverbeds.
They are found in association with lava fields, which is not particularly
surprising, as the rilles are now known to be lava tubes whose roofs have
caved in, exposing the streamlike courses followed by the molten lava

FIGURE 4.6. An oblique view taken by the Apollo 16 astronauts, looking south over the great craters Ptolemaeus (148 kilometers across, partially shown at the bottom of the image), Alphonsus (129 kilometers across), and Arzachel (97 kilometers across). Alphonsus is a pre-Imbrium impact crater whose floor is marked by a complex rille system and several volcanic cinder cones, which are the small craters surrounded by dark material. (NASA photograph)

during the period when lava was flowing from below onto the lunar surface.

The features called grabens, though superficially similar to rilles, are generally much straighter and represent areas of collapse along faults—points where local extensions of the crust have taken place. Other features associated with faulting are scarps, of which the best-known example is the Straight Wall (Rupes Recta) between the craters Thebit and Birt near the eastern border of Mare Nubium (fig. 4.8). It was first recorded by Huygens, but his observation was forgotten for more than a century until the feature was rediscovered by Schroeter. The Straight Wall casts a shadow before the time of full moon and appears then as a thin black line,

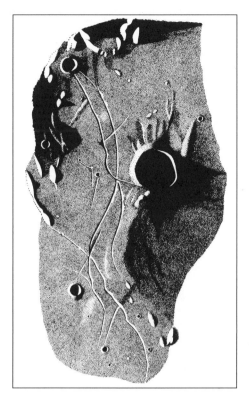

FIGURE 4.7. The Triesnecker rille system.
(Pen and ink drawing by Harold Hill, based
on observations made on Jan. 28, 1985,
18:25–19.15 UT, using a 10-inch
Newtonian reflector at 286×)

while after full moon its face is illuminated and appears bright. The height
of the drop-off is somewhat more than 200 meters, but the angle, at 40°, is
far from sheer.

Then there are the peculiar bubble-like swellings known as domes,
found in abundance in certain areas. There are, for instance, half a dozen
near the crater Hortensius just west of Copernicus. The typical dome is
less than 15 kilometers across, and its slopes are very shallow, on the order
of 5°. Some appear to be viscous lava deposits, while others have been
identified as small shield volcanoes, though they are minor examples
compared to the great shield volcanoes of the Earth or Mars.

As the Moon's interior cooled, lava flows became more and more

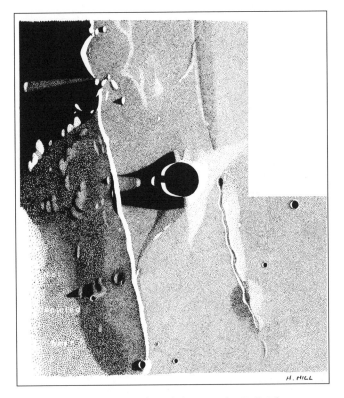

FIGURE 4.8. Sunset on Birt and the Straight Wall. The
Straight Wall is actually a fault, and here it shows its
characteristic sunset appearance, with its face brightly
illuminated. At sunrise the face is in shadow. To the right of
the Straight Wall is a sinuous rille at whose northern end (at
bottom in the image) there is an example of a lunar dome.
(Pen and ink drawing by Harold Hill, based on observations
made on Oct. 12, 1990, 03:20–04:45 UT, using a 10-inch
Newtonian reflector at 286×)

infrequent. By 3.2 billion years ago they were essentially over. The lunar
surface then looked very much as it does today. A few young features were
formed by subsequent impacts, of which outstanding examples are Copernicus, which is 810 million years old, and 84-kilometer-wide Tycho, 109
million years old, both centers of bright ray systems. The rays are made up
of finely pulverized ejecta materials, as was learned during the Apollo 12
mission in November 1969, which landed in Oceanus Procellarum on one
of the Copernican rays. Although not visible at all under conditions of low

illumination, at full Moon their dazzling radiations completely dominate the lunar scene.

Whence came the Moon in the first place? At the end of the nineteenth century, Sir George Darwin (Charles's son) suggested that the Moon formed through fission from the Earth. This was an attractive idea, but it had to be given up when it became clear that in order to have the proto-Earth spin fast enough to fission, so much angular momentum would have been transferred to the Moon that instead of going into orbit around the Earth it would have escaped completely, becoming an independent planet in orbit around the Sun. The alternative, of the Moon starting as an independent planet that was somehow captured by the Earth, also leads to serious difficulties, inasmuch as just the right velocity of approach is needed to avoid having the Moon simply sweep past the Earth or smash into it outright.

But suppose that a small planet about the size of Mars *did* make a grazing swipe at the Earth some 4.5 billion years ago. It would have broken apart, and its iron core would have been separated from its rocky mantle. The iron core would have dropped toward the Earth, sinking into the Earth's interior to become part of its iron core and leaving behind a Saturn-like ring of rocky debris from the disrupted planet's mantle to reaccrete into the Moon. This is the currently accepted theory, and at least it avoids the Achilles' heel of previous concepts by allowing just the right amount of angular momentum to be transferred.

In the early history of the Moon, the hemisphere now facing us must have been at the Moon's leading edge, smashing head-on into the ring of debris as the Moon circled the Earth, thus forming the large impact basins. Only much later, as the Moon shifted its orientation to correspond to its new balance point, did the hemisphere with the large impact basins come to face the Earth.

These scenes of violent collision, some involving whole worlds, stir the imagination and remind us of the stories of the chaos, or primordial abyss, out of which, in the myths of earlier peoples, the universe was supposed to have been summoned. In the lunar desolation, the features of the remote past have been preserved, hermetically sealed from the onslaught of such erosive forces as air and water, which have long since obliterated most of them from the Earth's surface. In crater-saturated highlands, lava-filled basins, and rilles, scarps, and domes, one may cite chapter and verse of the genesis of the Solar System.

Today, it is almost hard to believe that between July 1969, when Neil

Armstrong and Edwin Aldrin landed at Tranquillity Base, and December 1972, when the last of the Apollo missions came to an end, twelve men walked on the Moon. Even the last of the Russian probes, Luna 24, landed there as long ago as August 1976. As an object of scientific inquiry, then, the Moon has become rather passé. Many an amateur astronomer has, I suspect, been guilty of the same neglect, yet nowhere else is there so much to observe. The Moon is a geological wonderland, and anyone who is serious about understanding the surfaces of the other worlds of the Solar System will not return to it once but time and time again.

CHAPTER FIVE

Mercury

W e turn now from the Moon to a world that resembles it in many respects—the innermost planet, Mercury. Elusive in the glow of twilight as it races swiftly before or after the Sun, from which it never strays by more than 28°, Mercury is a brief visitor in our skies. Even at its highest, it hangs rather low in the dawn and twilight sky, so its light can only be viewed through the greatest depth of the Earth's atmosphere. Indeed, among the ancient Greeks its unplanetlike twinkling at such times earned for it the name Stilbon, the Scintillating One.

It used to be said that in all his life Copernicus never caught sight of the planet, owing to the mists rising from the Frisches Haff near Frauenburg, but now we know that this was untrue, and in any case the planet is easy enough to see provided that one looks for it when it is suitably placed in a reasonably clear horizon. At the opposite extreme from the Copernicus story is that of Gallet, a seventeenth-century cleric at Avignon who was nicknamed the Hermophile because he managed to see the planet more than a hundred times during his lifetime. Though Gallet's record may be good for conditions in southern France, that figure has no doubt been surpassed by many observers in better climates.

The best times to look for Mercury are when its angular distance from the Sun is greatest, that is, around the dates of its greatest elongations east and west of the Sun, when it appears as an evening and a morning "star," respectively. An average of six such elongations occur each year. For northern observers the best of them for seeing Mercury occurs in the evening in spring and in the morning in autumn.

Telescopically, Mercury shows phases. Galileo looked for them but was unable to see them owing to the small size of his glass. Apparently they were first recorded by an Italian Jesuit, Zupus, in 1639, and then by Hevelius five years later. A good 2- or 3-inch telescope will show them.

According to the Copernican theory, the Earth is an ordinary planet in orbit around the Sun, and thus it was logical for early observers to regard the rest of the planets as other Earths, and they—or at least the Copernicans among them—took for granted that the planets were all inhabited. A French writer, Bernard de Fontenelle, went so far as to suggest that, because of Mercury's proximity to the Sun, its inhabitants must be "so full of Fire, that they are absolutely mad; I fancy they have no Memory at all . . . make no reflections, and what they do is by sudden starts, and perfect haphazard; in short, Mercury is the Bedlam of the Universe." It is true that Fontenelle was not an astronomical observer himself, but his views do give some idea of what was thought at the time.

Not long after Fontenelle's *Conversations on the Plurality of Worlds* appeared in 1686, an observation was made that seemed to provide definite support for the idea of the planet's habitability. In 1707, John Flamsteed, observing at Greenwich, noted a fuzzy ring around Mercury during its transit across the Sun. This showed, he wrote, that the planet was "encompassed by a thick haze or atmosphere." The ring was also seen by a number of later observers, including Johann Schroeter during the transit of 1799 (its texture was, he said, a "mere thought"), but now we know that the planet is to all intents and purposes airless. The ring was only an illusion due to contrast (see fig. 5.1).

Schroeter also made a number of observations of Mercury in the twilight sky and at times saw the southern horn of the crescent as blunted, which he interpreted as being due to the roughness of the terrain there. Moreover, he detected a scintillating point just inside the unilluminated sector, which he identified as a lofty peak. From these observations he put Mercury's rotation period at close to 24 hours. Schroeter's scintillating point has long since been explained away as an effect of poor atmospheric seeing, yet for want of anything better his rotation period remained the standard for decades. Not until the 1880s was the problem tackled from a fresh viewpoint.

Enter Giovanni Virginio Schiaparelli, a keen-eyed and ingenious astronomer. In 1881 Schiaparelli began a long series of observations of Mercury with the 8.6-inch refractor of the Brera Observatory in Milan. Unlike previous observers, who had struggled with bad air during the twilight periods, Schiaparelli decided to try to observe the planet in broad daylight, when it was higher above the horizon (as his telescope was equipped

FIGURE 5.1. Transit of Mercury, November 1973, during its egress from the Sun. Note the bright ring around the planet's disk, which was once regarded as evidence of a dense atmosphere but which is now known to be an optical illusion. (Drawing by Richard Baum at Chester, England, based on observations made on Nov. 10, 1973, 13:15–13:16 UT, using a 4 ½-inch refractor at 186×)

with setting circles, which allow the telescope to be pointed toward a specific set of coordinates in the sky, he had no great difficulty locating the planet at such times). Under these conditions, Schiaparelli was able to make out a pattern of delicate sepia-colored streaks on the planet's surface (fig. 5.2), and at the planet's evening phase he noted a grouping of features that looked for all the world like the Arabic numeral 5.

Schiaparelli carried out some of his observations near superior conjunction, when Mercury was only 3° from the Sun. At such times the disk is extremely small, only 4 or 5 seconds of arc, but to make up for this, most of Mercury's sunlit side is then turned toward the Earth. By 1889 Schiaparelli had assembled 150 drawings of Mercury. The markings were less distinct at certain times than at others—this was especially so in the case of the feature that resembled a 5. Thus he was led to conclude that Mercury had an atmosphere dense enough to give rise to frequent clouds. He suggested, moreover, that the dark areas were probably similar to the maria of the Moon, but he added, "if anyone, taking into account the fact that there exists an atmosphere upon Mercury capable of condensation and perhaps also of precipitation, should hold the opinion that there was

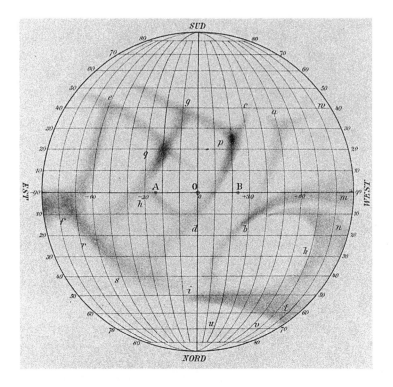

FIGURE 5.2. Giovanni Schiaparelli's 1889 map of Mercury. The map is oriented with south at the top, which corresponds to the view in the telescope. (From "Sulla rotazione di Mercurio," *Astronomische Nachrichten*, no. 2944 (1889).

something in those dark spots analogous to our seas, I do not think that a conclusive argument to the contrary could be given." The main finding from Schiaparelli's study, however, was that the planet's rotation period was much longer than Schroeter's 24 hours. In fact, Schiaparelli announced in 1889 that it was equal to the period of revolution—88 Earth days. As noted in chapter 4, the Moon presents a similar case of captured rotation with respect to the Earth, a result of the tidal forces of the Earth acting on the Moon. Presumably, or so Schiaparelli's argument went, Mercury's original period of rotation must have been slowed and finally halted by the far more powerful tidal forces of the Sun.

All of this seemed completely logical, of course, and Schiaparelli's 88-day rotation period was immediately accepted. A number of astronomers who took up the study of the planet confirmed his results—Henri Perrotin, Percival Lowell, René Jarry-Desloges, and Henry McEwen, for example. But the most skillful and influential of all was Eugène Michael Antoniadi, a native of Constantinople (now Istanbul) who spent most of his adult life in France. In the 1920s Antoniadi began a series of observations of Mercury with the 32 3/4-inch refractor at Meudon, Europe's largest. Not only did he support Schiaparelli's rotation period, he endorsed the clouds as well, and he produced a chart (fig. 5.3) that remained the standard until 1965. On it Antoniadi proposed names for the various features. There was, for instance, a large dark area on the southeastern part of the disk, which he named Solitudo Hermae Trismegisti (the Wilderness of Hermes the Thrice Greatest). On the western part of the disk were several spots corresponding to Schiaparelli's figure 5, of which the main ones received the names Solitudo Atlantis, Solitudo Criophori, and Solitudo Aphrodites.

Antoniadi could hardly have been more confident of his results, and at the time it seemed that a good deal was known about the planet—even something of the atmospheric circulation patterns. For example, in 1935 he wrote:

> Schiaparelli had remarked, with wonderful keenness (and it is difficult adequately to admire his work on Mercury) that the aerial veils of that orb are more frequent on the evening than on the morning phase, more frequent over the combination of dusky spots forming his number 5 than on any other region. The writer was enabled to confirm in detail this statement: *Solitudo Criophori* on the accompanying chart, often rendered invisible by the interposition of local cloud, appeared really much more frequently dimmed by these veils than any other greyish mark, the percentage of cloud over it amounting to about 45, while it curiously stood only as low as 5 over the greatest of the dusky areas, the half-tone *Solitudo Hermae Trismegisti,* whose size approaches that of Australia.

There was no doubting Antoniadi's ability as an observer. Nevertheless, his clouds were always viewed somewhat circumspectly. At 4,878 kilometers in diameter, Mercury is only half again as large as the Moon, and its mass is only one-eighteenth that of the Earth. This means that it has a weak gravitational pull, and because of the planet's proximity to the Sun, any gas molecules there are bound to be heated into very violent motion (the surface temperature at perihelion reaches 430°C). The molecules would, then, have every chance of escaping, and it was difficult to see how such a planet could ever retain an appreciable mantle of air.

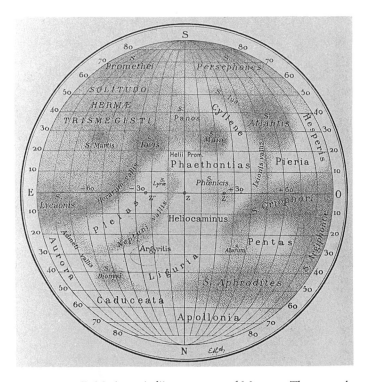

FIGURE 5.3. E. M. Antoniadi's 1933 map of Mercury. The general
appearance is much like that of Schiaparelli's 1889 map. (From
E. M. Antoniadi, "La Planète Mercure: Sa Géographie, sa Rotation
et ses Voiles Atmosphériques," *Bulletin Société Astronomique de
France* 47 (1933): 545)

Later observers, such as Bernard Lyot and Audouin Dollfus of the Pic
du Midi Observatory, located high in the French Pyrenees, were equally
convinced of the captured rotation, and in the early 1960s Dollfus went so
far as to write that Mercury's rotation had been shown to agree with its
period of revolution to within one part in ten thousand. If there was one
datum of planetary astronomy that seemed beyond doubt, it was this.
Then in 1965 there came a shock from the radio astronomers.

What our eyes perceive as visible light is only a narrow part of the entire
electromagnetic spectrum. Beyond the violet end we have the invisible
ultraviolet, X-ray, and gamma-ray regions, while the red shades into the
infrared, microwave, and radio regions. Radio waves have the longest
wavelengths and can be detected using a radio telescope.

A radio telescope can also be used to send out a pulse, the reflection of which, say from a planet's surface, can be detected. In this case the radio telescope is being used as both a transmitter and a receiver. Radar echoes from the Moon were detected as far back as 1947, and by 1965 the technology had improved so much that Mercury was within range. If the echo is from the surface of a rotating planet, the approaching limb will reflect waves shifted to a slightly higher frequency, and the receding limb to a slightly lower frequency, an example of the well-known Doppler effect. From exact measurements of these shifts, the rate of the planet's spin can be determined, and in this way radio astronomers discovered that Mercury rotates once in 58.65 Earth days. This is exactly two-thirds of the planet's period of revolution around the Sun. How could the visual observers have been so deceived?

It turns out that 58.65 days is not only two-thirds of Mercury's period of revolution around the Sun, it is also close to half of the period between its successive appearances in the same phase as viewed from the Earth (the synodic period, in Mercury's case 116 days). Thus, as Mercury came to its greatest elongations from the Sun and was best placed for observation, astronomers saw the same features and concluded that Mercury always kept the same face toward the Sun. The selective observation of the planet through periodic observing windows has been called the stroboscope effect. The synchronisms in the periods involved are not quite perfect, and after about seven years new regions of the planet begin to swing into view. But the idea of Mercury's marked libration, the belief in a cloudy atmosphere around the planet, and above all, preconception—the tendency of observers to convince themselves, when looking at markings near the threshold of visibility, that they were seeing the same markings as those seen on other occasions—conspired to keep this recognition from Schiaparelli, Antoniadi, and the rest.

Once the true rotation of Mercury was discovered, it was no longer necessary to believe in the existence of clouds, and this is just as well. Whenever a known dusky patch had failed to show itself even when conditions were favorable, the visual observers, who were committed to the idea of captured rotation, had no choice but to conclude that it was being screened by atmospheric mists. In fact, Mercury does have an extremely tenuous atmosphere, which consists largely of sodium atoms and also, apparently, water vapor; see below. The pressure is only about one trillionth that of the Earth.

Mercury's somewhat bizarre rotation is, according to a theory put forward by Peter Goldreich, a result of spin-orbit coupling. At one time, Mercury probably had a fast rotation. However, as the planet spun,

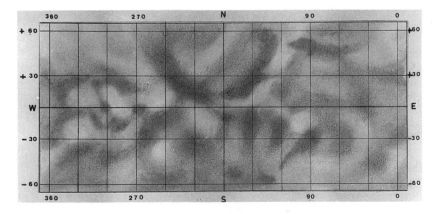

FIGURE 5.4. This map of Mercury was drawn by the skilled French observer Audouin Dollfus in 1972, after the radar discovery of the planet's true rotation period. Some of the features shown on the maps by Schiaparelli and Antoniadi can be identified. Dollfus's map is based on observations with the 24-inch refractor at the Pic du Midi Observatory in France. (Courtesy Audouin Dollfus)

centrifugal force would have produced an equatorial bulge, and given that the planet travels in a highly eccentric orbit—its distance from the Sun ranges from 46 million kilometers at perihelion to 69.8 million kilometers at aphelion—the Sun would have pulled most strongly on this bulge when the planet was near perihelion. Eventually tidal friction would have slowed the rotation so that the long axis pointed exactly toward the Sun whenever the planet came to perihelion.

Goldreich's theory is obviously a modification of the old idea of a 1:1 captured rotation produced by tidal forces, and indeed the 3:2 ratio of revolution to rotation represents another dynamically stable configuration. Yet not only does the same axis of Mercury's globe point toward the Sun at each perihelion passage, but the axis is also aligned with the Earth whenever Mercury comes to inferior conjunction. Thus Mercury appears to be in spin-orbit coupling not only with the Sun but with the Earth—a surprising result insofar as the tidal force of the Sun on Mercury is always a millionfold greater than that of the Earth on Mercury.

Because of Mercury's 3:2 spin-orbit lock, alternate faces of Mercury are turned toward the Sun each time the planet passes perihelion. The subsolar points obviously receive the most intense heat and are known as hot poles. The temperature there can reach 430°C—high enough to melt lead—but during the night it plummets to −185°C, or just above the point where air begins to liquefy. Incidentally, one of the hot spots is taken as defining 0° of longitude on Mercury; the other lies at 180°.

A day on Mercury would be rather strange. At one of the hot spots, the Sun would rise in the east, climb into the sky for six weeks, then make a retrograde loop for eight days before continuing westward and finally setting six weeks later. At longitudes 90° and 270°, the retrograde loop would be observed when the Sun is at the horizon; thus observers there would witness double sunrises and sunsets. The retrograde loop occurs because, for eight days near perihelion, Mercury's velocity in its orbit exceeds its rate of spin on its axis.

Mercury has been relatively neglected by interplanetary spacecraft. Only a single craft, Mariner 10, has so far paid a visit to the planet. Launched on November 3, 1973, it passed within 5,670 kilometers of Venus on February 5, 1974, then headed to a first encounter with Mercury on March 29, 1974. By making several corrections to the spacecraft's trajectory, it was possible to bring it back to Mercury for two additional encounters, which occurred on September 21, 1974, and March 16, 1975. The closest approach came during the third flyby, when the spacecraft glided to within 327 kilometers of the planet's night side.

The surface of Mercury as revealed by Mariner 10 is at least superficially similar to the Moon's. Mariner 10 photographed only about half of the total surface area, but since then, using Earth-based radar, the rest of the planet has been roughly mapped. It turns out that about 40 percent of the Mariner area and 20 percent of the radar area are covered with smooth plains. Some of the plains contain fresh craters with bright systems of rays. An especially exciting, and unexpected, discovery, made from radar studies of the planet in August 1991, is of a water-ice cap at the north pole. There is every reason to believe that there is another such cap at the south pole. Because Mercury's axis is nearly perpendicular to the plane of its orbit, the sunlight at the poles arrives very obliquely, and the temperature at the poles remains cold enough for gases in the thin atmosphere to precipitate out as snow.

To facilitate comparison with classical maps, I might add that the Mariner 10 zone of coverage (between longitudes 10° and 190°) corresponds to the Solitudo Hermae Trismegisti part of the planet, while Solitudo Criophori and the rest of Schiaparelli's figure 5 formation lay in darkness just beyond the terminator. On the whole, the location of the smooth plains on Mercury agrees as well as can be expected with the dark features drawn by Schiaparelli, Antoniadi, and others, and it may even be that a few observers glimpsed some of the other details. For example, W. F. Denning, using a 10-inch reflector on several mornings in November 1882, noted a tiny white spot of unusual brilliance from which luminous veins or radiations spread to embrace a wide area. It seems clear that Denning had glimpsed a crater and one of the bright ray systems.

The dominant feature on the surface of Mercury is the huge multi-ringed plain known as the Caloris basin, which lies near one of the hot poles of the planet (hence the name, which means Basin of Heat). It was captured right at the terminator in the Mariner 10 photographs, so it shows up in high relief (fig. 5.5). The Caloris basin is a feature of the same type as Mare Imbrium on the Moon and is identical to it in size, measuring 1,300 kilometers from rim to rim. So violent was the impact that formed it that it even shaped surface features on the other side of the planet—the peculiar hills and valleys there having formed, it is believed, by the convergence of seismic waves from the Caloris event.

Within the Caloris basin's mountainous rim—actually its impact ejecta blanket—the terrain is unlike anything found elsewhere in the Solar System. It consists of material that evidently filled the basin after its formation and then was fractured by strains in both radial and concentric directions when the basin first sank slightly, then subsequently rose. Beyond the rim are low ridges and hills interrupted by radial valleys, which give way in turn to smooth plains—the Suissei Planitia to the northeast, the Odin Planitia to the east, and the Tir Planitia to the southeast. Superimposed on the hilly terrain, and thus obviously formed after the Caloris basin, is the smooth-floored crater Mozart, 300 kilometers across.

Among other notable craters on Mercury are the flooded craters Beethoven and Tolstoi; the former is over 600 kilometers across. Centers of bright ray systems include Kuiper, Snorri, and Copley.

Owing to the greater surface gravity of Mercury, ejecta blankets surrounding impact features are not as extensive as those on the Moon. Scarps are especially common and provide evidence of a past contraction of the whole planet. This may well have been a result of the tidal slowing of its rotation mentioned above. As the planet came to spin less rapidly, its oblate shape would have tended more toward the spherical, collapsing inward upon itself and fracturing the crust. Another possibility is that the planet's shrinkage was related to the cooling of its core, but the result would have been the same.

Further evidence of global shrinkage is found in the differences between the lunar basins and the less well defined plains of Mercury. On the Moon,

FIGURE 5.5. The Caloris basin of Mercury can be
seen at the terminator, on the extreme left, in this
photograph taken by the Mariner 10 spacecraft
during its flyby in 1974. This is a well-preserved
multiringed basin comparable to Mare Orientale on
the Moon. (NASA photograph)

lava filled the basins during the half billion years or so after they formed.
On Mercury, the lava flows stopped much earlier, perhaps even before the
last period of heavy bombardment. Presumably this was because on
Mercury the general global compression of the crust cut off the conduits
carrying magma from the interior up to the surface.

There are other fundamental differences between Mercury and the
Moon. In composition the Moon is mostly rock, and it is not certain that it
has an iron core at all. Mercury, in contrast, has a comparatively thin
rocky shell surrounding an immense iron core—by volume, the planet is

40 percent iron. The discovery of this iron core was something of a surprise. Though planets nearer the Sun ought to be composed mostly of materials with high vaporization temperatures, since at such distances only these materials would have had a chance of condensing from the solar nebula, Mercury still has too much iron. It may well be that a collision between Mercury and another planet-sized body early in its history was responsible for tearing away the rocky mantle and leaving behind the present planet, with its disproportionate iron core.

Another surprise is that Mercury has a magnetic field. Its intensity is only 1 percent of the Earth's, but this is still highly significant. The standard theory used to be that both a molten core and a rapid rotation were necessary to produce a magnetic field. Thus Venus, Mars, and the Moon have none—Venus because its rotation is too slow, Mars because its core is cold, and the Moon for both reasons. However, the theory completely breaks down in the case of Mercury. The planet is clearly in the slow rotation camp, and its core is cold (as has recently been settled once and for all by observations of radio-wave emissions from the surface, which show that the planet as a whole is no warmer than expected from solar heating alone; this means that there is no internal heat source). Thus Mercury must generate its magnetic field through residual magnetism locked into its minerals when its core solidified.

For a while during the nineteenth century, astronomers believed that a planet existed nearer the Sun than Mercury. The announcement was made by Urbain Jean Joseph Leverrier of France, probably the most brilliant mathematical astronomer of his day.

Mercury's motion had long been a problem for astronomers. In particular, predictions of the planet's transits across the Sun had usually been wildly in error, and after reviewing the available transit observations going back to 1697, Leverrier found that the reason for this was that the perihelion of Mercury's orbit seemed to be precessing around the Sun at a rate of 38 seconds of arc per century so that it wound completely around the Sun in a period of three million years. Such a precession was completely unexpected, and Leverrier could only ascribe it to "some as yet unknown action."

Leverrier made this remarkable discovery in 1849. Three years earlier he had won renown for his discovery of the planet Neptune, whose position in the sky he had pinpointed by his masterful analysis of the discrepant motion of Uranus using the methods of Newtonian mechanics. The discovery of Neptune was, as Richard Baum has written, "an unparalleled triumph; quite literally the zenith of Newtonian mechanics. More importantly the collapse of that edifice had been averted by tools created within itself."

Leverrier coolly attempted a solution to the Mercury problem along similar lines. Thus in September 1859 he announced to the French Academy of Sciences that Mercury was being acted upon by an unknown mass lying closer to the Sun, possibly a small planet but more likely a swarm of intra-Mercurial asteroids or even a ring. Leverrier suggested that astronomers would do well to be vigilant of any unusual dark speck which might appear on the solar disk. But in fact they had already been doing so for many years. Thus Heinrich Schwabe, a German pharmacist and amateur astronomer, had been keeping careful records of the solar spots since 1826, initially with the purpose of detecting an intra-Mercurial planet. Though the planet failed to show itself, Schwabe was rewarded with a far more important discovery: the eleven-year sunspot cycle. An American, Edward Herrick, had also been unsuccessfully chasing such a planet since 1847, and there were others as well.

Less than three months after his announcement to the French Academy, Leverrier received a curious letter from a country doctor, Edmond Lescarbault, who practiced medicine in the French village of Orgères. Lescarbault was an amateur astronomer, and he claimed to have witnessed the transit of a small body across the Sun on March 26, 1859. Camille Flammarion, who knew Lescarbault personally, later described him as "my old and excellent friend" and as "one who devoted to admiration of the beauties of the sky the time which was not taken up in alleviation of earthly miseries." Leverrier had never heard of Lescarbault and had no such basis of trust. He was frankly skeptical, but his interest had been piqued, and he could not resist paying a visit to Orgères himself.

On December 30, 1859, Lescarbault answered a knock on his door only to be greeted brusquely by a stranger: "It is you, then, sir, who pretend to have observed the intra-Mercurial planet, and who have committed the grave offense of keeping your observation secret for nine months. I warn you that I have come here with the intention of doing justice to your pretensions, and of demonstrating either that you have been dishonest or deceived. Tell me, then, unequivocally what you have seen." The stranger was, of course, Leverrier, though it is doubtful whether he ever identified himself during his visit. Lescarbault repeated his account of what he had seen and showed Leverrier his timepiece (a silk pendulum used to beat out seconds), and his telescope. Next Leverrier asked to examine the record of Lescarbault's observation, and Lescarbault produced a laudanum-stained paper that had been serving as a marker in a volume of the French ephemeris *Connaisance des Temps*. What distance had Lescarbault derived for the planet? Lescarbault confessed that, despite many attempts, he was no mathematician and had been unable to solve the problem. Still Leverrier pressed him, demanding to see the rough drafts of these calculations. Lescarbault replied, "My rough drafts! Paper is rather scarce with us. I am a joiner as well as an astronomer. I calculate in my workshop, and I write upon the boards; and when I wish to use them in new calculations, I remove the old ones by planing."

By the end of the interview, which lasted for about an hour, Leverrier's attitude had softened toward the country doctor, and he left Orgères convinced that Lescarbault had actually seen an intra-Mercurial planet. The year 1859 was an important one for science. It saw the publication of *The Origin of Species* in England, the announcement of the spectroscopic laws by Kirchoff and Bunsen in Germany, and—so it seemed—the discovery of a new planet in France. Leverrier was credited with having predicted the existence not only of the outermost planet in the Solar System but of the innermost as well, and for his part, Lescarbault was named a member of the Legion of Honor.

Leverrier wasted no time in calculating an orbit for the new planet, and it was even given a name—Vulcan. Several observations that came to light seemed to record earlier sightings, and forthcoming transits were predicted for March 29, April 2, and April 7, 1860. Unfortunately, scrutiny of the Sun's disk at the expected times revealed nothing unusual, and special searches conducted during the total solar eclipse of July 18, 1860, were equally unsuccessful.

Indeed, aside from several disputed observations, most of which, on closer examination, could be discounted as ordinary small sunspots, the planet was never seen again—though Leverrier continued to believe in its existence right up to his death in 1877. Thirty years after Lescarbault's supposed discovery, Camille Flammarion summed up the attitude of growing skepticism. For thirty years, he wrote, "there has not passed a single day, so to say, without the Sun being examined, drawn, or photographed in Italy, England, Portugal, Spain, America, France, and elsewhere; that the supposed planet would have passed more than a hundred times across the Sun, and that, nevertheless, it has never been seen means that it is either well hidden, or it does not exist. Mercury was the god of thieves; his companion steals away like an anonymous assassin!"

But if Vulcan did not exist, the problem it had been conjured up to solve remained. Historian of astronomy Agnes M. Clerke described the precession of Mercury's perihelion in 1893 as "one of the pending problems of astronomy," and the American astronomer Asaph Hall went so far as to propose a slight correction in Newton's inverse square law of gravitation. Here too there were analogies to the Uranus problem, where modifications of the inverse square law had been contemplated, only to be abandoned when Neptune was discovered. But in the case of the precession of Mercury's perihelion, the problem never was to be solved from within the Newtonian framework—though of course there was no way Leverrier or anyone else in the nineteenth century could have known this.

The problem was finally solved in November 1915. At that time Albert Einstein succeeded in deriving the observed precession of Mercury's perihelion from his General Theory of Relativity, which introduced a new theory of gravitation to replace that of Newton. It was one of the most emotional experiences of his life. "Imagine my joy . . . at the result that the equations yield the correct perihelion motion of Mercury," he wrote to a friend in January 1916. "For a few days, I was beside myself with joyous excitement." Clearly there was no longer any need to believe in Vulcan, yet the question remained: What had Lescarbault seen?

Perhaps nothing. It seems that a skillful French astronomer, Emmanuel Liais, had been observing the Sun from Brazil on March 26, 1859, at the exact moment of the alleged transit, and he saw nothing unusual. Liais denounced Lescarbault as a fraud, though he later changed his mind and conceded the doctor's integrity and good faith. Unfortunately, this does nothing to explain his observation, and we may never know for certain just what it was he saw. But at least there can no longer be any doubt that Mercury is the innermost of all the worlds that whirl around the Sun, with the exception of the Sun-grazing comets and the rare asteroids, like Icarus and Phaethon, which dart still nearer to its searing rays.

CHAPTER SIX

Venus

Seen now before dawn, now in the evening gloaming is a lovely "star" more brilliant than any other in the heavens. The Egyptians gave it two names, Ouâti when it appeared in the evening sky, Tioumoutiri when it appeared in the morning. So did the early Greeks, who referred to it as Hesper (West) and Phosphor (Light Bearer).

Because of its remarkable beauty, the planet was generally worshipped as a female divinity. By the sixth century B.C. the fact that Hesper and Phosphor were one body had been established, and the later Greeks referred to it as Aphrodite, in honor of their goddess of love and beauty—the same goddess later worshipped by the Romans as Venus. Among the Babylonians it was Ishtar, sister of the Moon, and the Assyrians knew it as Ashtaroth or Astarte.

The English poet John Milton, in his poem "The Hymn on the Morning of Christ's Nativity," written about 1629, recalls

mooned Ashtaroth,
Heav'n's Queen and Mother both.

The adjective "mooned" no doubt alludes to the phases through which the planet passes, presumably unknown before the invention of the telescope. Milton knew of them from Galileo's recent telescopic observations, but the association of the goddess with the crescent horns was not original with Milton but rather goes back to antiquity. A figure of Astarte bearing a crescent-tipped staff has been found among the ruins at Nineveh.

This raises the interesting question of whether the ancients could possibly have known that the planet sometimes shows a crescent phase. Antonie Pannekoek, in his *History of Astronomy,* suggests that the Babylonians may have done so: "It does not seem impossible that in the clear atmosphere of these lands the horns of the Venus crescent were perceived; modern astronomers, too, have mentioned such instances. An American missionary, D. T. Stoddaert, in a letter to John Herschel from Oroomisha in Persia in 1852, wrote that at twilight Jupiter's satellites and the elongated shape of Saturn could be seen with the naked eye and that through a dark glass the half-moon shape of Venus immediately struck the eye. . . . Thus it can be still better understood that in these ancient times the Babylonian priest-astronomers devoted special attention to Ishtar as a sister-star to the Moon."

All of this sounds plausible enough, but I am skeptical. In June 1988, conditions were especially favorable for making a test, as the planet passed 8° north of the Sun at inferior conjunction and thus could be followed with the naked eye to within only a day or two of conjunction, when its apparent diameter was a full minute of arc. I made a point of carefully watching the planet about this time. Certainly the possibility of detecting the sickle shape was entirely out of the question, and all I can say is that even detecting a slight elongation was well beyond my means, at least under average atmospheric conditions.

At any rate, the official discovery of the phases of Venus was made by Galileo with his small telescope in 1610–11. According to Ptolemy, Venus could never be seen as more than a crescent; in the Copernican system it could appear not only as a crescent but as a half or even a gibbous phase. This was, then, the real *crux observationis,* and Galileo's observations showed decisively that the planet passes through the full gamut of phases, dealing the Ptolemaic theory a blow from which it never recovered.

Venus's spectacular appearance as seen with the naked eye—it becomes bright enough to cast a shadow, and it is visible in broad daylight—does not prepare one for its disappointing telescopic performance. One can see the phase but usually little else. The planet's brilliance is partly due to its nearness to the Earth (it comes to within 42 million kilometers at inferior conjunction) and partly to its being covered with highly reflective clouds, which throw 80 percent of the incoming sunlight back into space. Dazzling as they are, these clouds appear nearly featureless to the visual observer. Usually there are only a few vague shadings at best (see fig. 6.1), and even such a basic question as the planet's rotation was long a matter of debate.

The color, incidentally, is usually described as creamy white. Actually, however, the planet's clouds are almost as strongly yellowish as those of

FIGURE 6.1. Venus, 1961, from observations by Richard Baum with a 4 ½-inch refractor at 186×. These impressions, by an observer of exceptional skill, show the nebulous dark shadings and the bright cap on the south cusp (at the top of the image here). (Left: March 16, 1961, 18:30 UT; right: March 19, 1961, 18:35 UT)

Jupiter. They appear white only because the planet's overwhelming brilliance completely saturates all three of the eye's color receptors, red, green, and blue. The brain combines this information to perceive white.

G. D. Cassini, at Bologna, managed to make out a few spots in 1666–67, and though his account of them is confusing in some respects, it seems that he concluded that the rotation period was somewhat less than 24 hours. Cassini's son Jacques later published a revised period based on his father's observations of 23 hours, 20 minutes, a period long quoted in the secondary literature as that of G. D. Cassini himself.

After moving to Paris in 1669 to take up the directorship of the observatory founded by Louis XIV, Cassini never recovered the spots. The next astronomer to report success was Francesco Bianchini, papal chamberlain at Rome, who produced a number of sketches in 1726–27. The spots appeared only along the terminator of the planet, and though Bianchini admitted that they were far from easy to see, he considered them definite enough to fix the rotation at 24 days, 8 hours. Moreover, he went so far as to draw up a map showing "oceans" and "continents," which was long taken more seriously by astronomers than it deserved to be.

Bianchini clearly believed that he was looking directly at the surface of

the planet, an impression that was shared by most of the pioneer observers. The first person to suggest that Venus is actually surrounded by a
dense atmosphere was Mikhail Lomonosov, a Russian, who came to this
conclusion after seeing a lucid ring around Venus during its transit across
the Sun in 1761. Unlike the ring seen around Mercury, which is purely an
optical effect, the ring around Venus is indeed produced by the planet's
atmosphere and remains visible even when the planet is off the Sun's disk.

Lomonosov's work did not become generally known for more than a
century, but meanwhile other observations supporting the idea that Venus
has an atmosphere were forthcoming. William Herschel, for instance,
reached this conclusion mainly from the planet's perpetually bland appearance, while Johann Schroeter called attention to the marked falling
off of light at the terminator, which he correctly explained as being due to
atmospheric absorption. Schroeter was also the first to observe extensions
of the horns of the crescent beyond a semicircle around the time of inferior
conjunction (see fig. 6.2), a Venusian twilight effect, and this furnished still
further evidence of an atmospheric envelope.

All of this was good, sound work. But in addition Schroeter published
more controversial findings. He described a slight deflection of the planet's
southern horn into the form of a hook, and on several occasions noted a
detached luminous point flickering in the darkness just off the cusp. From
these observations he concluded in 1792 that there were lofty mountains
near the south pole of the planet, and he calculated that his "enlightened
mountain" reached a height of 47 kilometers.

Herschel reacted to these reports with uncharacteristically sharp crit-
icism. It was, Schroeter wrote, such as he "could not reconcile . . . with the
friendly sentiments its author [had] always hitherto expressed toward
me." The mountains of Venus nevertheless remained standard Venusian
lore well into the next century. The French astronomer Leopold Trouvelot
wrote in 1878 that the bright polar areas of the planet appeared to him
"like a confused mass of luminous points, separated by comparatively
somber intervening spaces. This surface is undoubtedly very broken, and
resembles that of a mountainous district studded with numerous peaks, or
our polar regions with numerous ice-needles brilliantly reflecting the
sunshine." Ten years later, Camille Flammarion concurred: "Observations of the indentations visible on the crescent of Venus show that the
surface of the planet is quite as uneven as that of the Earth, and even more
so, and that there are Andes, Cordilleras, Alps, and Pyrenees."

Though we now know that there are indeed mountains on the planet's
surface, these have nothing to do with the mountains reported by the old

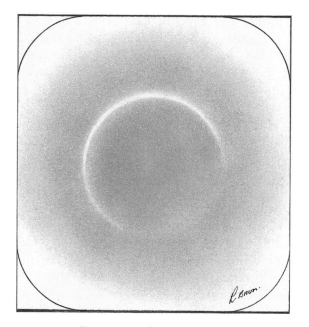

FIGURE 6.2. Venus near inferior conjunction. The prolongations of the cusps, which here reach all the way around the planet to form a complete circle, were first observed by Schroeter and are due to twilight effects in the planet's atmosphere. (Drawing by Richard Baum based on observations made at Chester, England, on April 2, 1977, 18:40 UT, using a 4 ½-inch refractor at 186×)

visual observers. Admittedly, the impression of irregularities in the terminator can be striking at times, and at least some of them are genuine enough, but they are due to clouds rather than to mountains. On the other hand, the "corrugated" terminator sometimes reported is entirely spurious and is explained by rippling of the image due to currents in the Earth's atmosphere.

From his "enlightened mountain" observations, Schroeter worked out a rotation period of 23 hours, 21 minutes, 19 seconds, which he later revised to 23 hours, 21 minutes, 7.9 seconds. This stood until 1841, when the Italian astronomer Father Francesco De Vico "corrected" it to 23 hours, 21 minutes, 21.9345 seconds. (One can only say that the decimal points, at least, were a bit optimistic.) Such a rotation period, close to the terrestrial one, lent credence to the view that Venus was a near-identical twin to the Earth. Thus Camille Flammarion wrote in 1880:

> Of what nature are the inhabitants of Venus? . . . All that we can say is, that organized life on Venus must be little different from terrestrial life, and that this world is one of those which resembles our own most. We will not ask, then, with the good Father Kircher, whether the water of that world be good for baptizing, or whether the wine would be fit for the sacrifice of the Mass; nor, with Huygens, whether the musical instruments of Venus resemble the harp or the flute; nor, with Swedenborg, whether the young girls walk about without clothing. . . . The only scientific conclusion which we can draw from astronomical observation is, that this world differs little from ours in volume, in weight, in density, and in the duration of its days and nights. . . . It should, then, be inhabited by vegetable, animal and human races but little different from those which people our planet.

Probably most astronomers of the day would have agreed with Flammarion's assessment. Then in 1890 G. V. Schiaparelli dropped an astronomical bombshell. Rejecting all previous work on the rotation as untrustworthy, the Italian astronomer based his conclusions entirely on his own observations with the 8.6-inch refractor in Milan. Only once in many years of work at the telescope had he seen anything definite on the planet: in December 1877, he had noted two bright oval spots near the south pole, which were unusually conspicuous by Venusian standards and unusually long-lived—he kept them under observation for a full month. Moreover, since he observed the planet in broad daylight rather than in the brief twilight periods, Schiaparelli was able to follow the spots for up to eight hours at a time, but not once did he detect any movement relative to the terminator. It followed that the rotation had to be very long—six to nine months, he concluded, and most likely in 1:1 spin-orbit lock with the planet's period of revolution, 225 days.

Over the next few years some astronomers published confirmations, while others continued to defend the short rotation. Unlike Schiaparelli's announcement of a 1:1 spin-orbit lock for the rotation of Mercury, the rotation of Venus remained unsettled, and in fact the visual observers were never able to solve the problem. Only in 1962 did it yield, finally, to the methods of radio astronomy. That year, radio astronomers announced that Venus's rotation is in a retrograde direction—that is, east-to-west rather than west-to-east like the Earth's rotation. Since the planet thus rotates "backward," the sidereal day, or the period of rotation as measured with respect to the stars, is, at 243 days, longer than the 225-day Venusian year. Moreover, the Sun rises in the west and 59 days later sets in the east.

Curiously, the rotation of Venus seems to be, like that of Mercury, in spin-orbit coupling with the Earth. Every time Venus comes to inferior conjunction and forms a line with the Earth and Sun, the same side faces the Earth. True, the resonance is not quite exact—the rotation period is 243.01 Earth days, compared with the resonance period of 243.16 days—but it can hardly be a coincidence.

The atmospheric features show a markedly different rotation, which was first recognized by Charles Boyer, a French amateur living in Brazzaville, the Congo, in the 1950s. When Boyer began his study, it had already been known for thirty years that Venus, nearly blank in visual wavelengths, shows definite patches in the ultraviolet. Using only a 10-inch reflector, Boyer obtained a series of photographs of these patches, on which he noted that the same configurations—in particular, a distinctive Y-shaped marking along the equator—tended to recur at roughly four-day intervals. He concluded, naturally enough at the time, that this represented the true rotation period of the planet.

After the radar results of 1962 were announced, astronomers realized that Boyer's four-day period actually corresponded to the circulation of the planet's atmosphere. This conclusion was confirmed in 1974 by the Mariner 10 spacecraft, which photographed Venus in the ultraviolet before heading on to Mercury. The Mariner 10 photographs showed the dark markings—later found to be clouds rich in sulfur compounds, which are good ultraviolet absorbers—in exquisite detail and nicely demonstrated the four-day circulation. Similar results were obtained by the Pioneer Venus probe a few years later (see fig. 6.3). The circulation pattern is driven by winds blowing at 400 km/hr, which means that the atmosphere of Venus rotates at a rate some sixty times that of the planet itself.

It is not, incidentally, the absolute speeds of the Venusian winds that are so remarkable—after all, on the Earth the jet streams embedded in the mid-latitude westerlies blow at well over 500 km/hr. Rather, it is the high speed relative to the very sluggish rotation of the solid body of the planet that is significant. On the Earth, the atmospheric "superrotation" of the mid-latitude westerlies is balanced out to within a few percent by the "subrotating" tropical easterlies. Thus the net angular momentum of the Earth's atmosphere comes out to within a few percent of that expected for

FIGURE 6.3. Ultraviolet images of Venus obtained by the Pioneer Venus spacecraft in February 1979, showing the rapid rotation of cloud features in the planet's upper atmosphere. Note the bright cloud swirls about the poles, which correspond to the "cusp caps" seen by visual telescopic observers. (NASA photograph)

an atmosphere co-rotating with the surface. On Venus, the "superrotating" equatorial region carries an excess of angular momentum of some tenfold—a state of affairs that has yet to be totally explained.

In addition to the dusky markings already described, other telescopic features worth mentioning are the bright cloud caps about the poles, first recorded by Gruithuisen in 1813. Some observers, most notably E. M. Antoniadi, held that the caps were mere contrast effects, but they are now known to be real—they correspond to bright cloud swirls about the planet's poles. Finally there is the strange glowing seen from time to time on the planet's night side—the so-called ashen light. The first report of the ashen light goes back all the way to Riccioli, who in 1643 described a coppery glow on Venus's dark side superficially resembling the phenomenon of the "Old Moon in the New Moon's arms" seen whenever the Moon appears as a slender crescent. In the case of the Moon this is, of course, due to the reflection of sunlight from the Earth, then nearly full as seen from the Moon, onto the lunar dark side. But Venus has no satellite, and thus there is no reason to expect anything of the sort.

The Earth itself would appear as a splendid object from Venus. When the separation between them is least, Venus is at inferior conjunction as seen from the Earth and has its dark side toward us, but as seen from Venus the Earth is at opposition and shows its full face. Thus the Earth would reach a magnitude of -6.5, compared with -4.5, which is the brightest Venus ever appears in our skies. But even so, Earthshine on Venus would be only 1/12,000th of what it is on the Moon, so the ashen light, if real, must be due to another cause. The answer may well be lightning storms on Venus's night side, for which there is at least indirect evidence from spacecraft observations. Whatever it is, it does seem to be a real phenomenon, not an illusion.

Lush as Venus appears from a distance, beneath its lovely clouds the surface is infernally hot. The temperature reaches 477°C (890°F) in the lowest-lying areas, well above the melting point of lead. Even on the highest peaks it is still 377°C (710°F). How in the world did Venus get so hot? The explanation involves the well-known greenhouse effect. In a greenhouse the glass is transparent to visible light. This is absorbed by the ground and then reradiated in the invisible infrared. Since glass is not transparent to the infrared, the heat remains trapped.

In the Earth's atmosphere, gases such as carbon dioxide, methane, nitrous oxide, and water vapor play the role of the glass. These molecules vibrate, and their vibrations absorb energy in the infrared. Thus they act to trap heat radiated from the ground. Though present in very small quantities—carbon dioxide, for example, represents only 0.0003 percent of the Earth's total atmospheric mass—these molecules play the main role in determining the Earth's temperature, and thus very modest changes in their concentrations can have enormous effects.

Carbon dioxide is the main byproduct of the combustion of fossil fuels, and since the last century a measurable increase in atmospheric carbon dioxide has occurred as a result of human activities on the planet. Other greenhouse gases have also increased, notably the synthetic chlorofluoro-carbons, or CFCs, widely used in refrigerators, air conditioners, and aero-sol cans. Not only are the CFCs even more efficient at trapping heat than carbon dioxide, they destroy ozone in the stratosphere, which is our main shield against deadly ultraviolet radiation from the Sun. Already they have been found to rip a hole each spring in the ozone layer over Antarctica.

Venus's atmosphere is much more massive than the Earth's. Moreover, it is almost entirely carbon dioxide—about 97 percent—with smaller amounts of sulfur dioxide, sulfuric acid, and water vapor. All of these are very good infrared absorbers. Indeed, the Venusian atmosphere is so deadly efficient at trapping heat that even though only 2.5 percent of the sunlight actually penetrates all the way through the clouds, the planet's surface is even hotter than Mercury's dayside at perihelion.

Because of their similar size and mass, the Earth and Venus used to be thought of as twins, but in fact they could hardly be more dissimilar. What was it that caused them to develop in such different directions? Venus is now bone dry; its atmosphere contains a mere whiff, 0.001 percent, of the water present in the Earth's oceans. However, at the beginning of the Solar System the Sun itself had only 70 to 75 percent of the luminosity it has now, and there may well have been oceans lapping Venusian shores. The Earth, on the other hand, ought to have frozen over under such a dim Sun, but it never did. Instead it has maintained its temperature within reasonable limits for billions of years. The explanation for this has to do with the so-called carbonate-silicate cycle.

On the Earth, carbon dioxide is dissolved in rainwater to form carbonic acid (H_2CO_3). When the acidified rainwater falls on rocks, calcium and bicarbonate ions are leached from calcium-silicate minerals. These ions run off into rivers and are then carried to the oceans, where they are taken up into the calcium carbonate skeletons of sea organisms such as phytoplankton. After they die, the organisms sink to the ocean floor, and their skeletons form sediments of calcium carbonate rocks (the limestones and dolomites). Thus, on the Earth most of the carbon dioxide becomes locked up in calcium carbonate rocks—indeed, at present some 300,000 times more carbon dioxide is trapped in these rocks than is found in the atmosphere.

Seafloor spreading forces the sedimentary rocks to the margins of continents, where oceanic plates meet lighter continental plates and sink beneath them. The sedimentary rocks are thrust to great depths below the surface, where conditions of tremendous heat and pressure exist, and calcium carbonate fuses with silicates to form quartz. At the same time, carbon dioxide is recycled back into the atmosphere by volcanic eruptions.

Consider now how the system works in practice. Suppose, first, that the

Earth's temperature abruptly drops for some reason—for example, as a result of an unusually intense period of volcanic activity, throwing dust into the atmosphere and thereby blocking the sunlight. At the cooler temperatures there will be less evaporation of water from the surface, less rain, and less erosion of rocks. Because less carbon dioxide is deposited in the rocks, more of it remains at large in the atmosphere. This excess, combined with the fairly constant contribution of carbon dioxide from volcanoes, leads to a net buildup of carbon dioxide in the atmosphere—and, through the greenhouse effect, to a rise in temperature, largely offsetting the initial fall.

Conversely, if one begins with a rise rather than a fall in global temperatures, the system would react in the opposite direction. There would be more evaporation of water, more rain, and more erosion of rocks, and thus more carbon dioxide would be deposited in the rocks. The point of all this is that the carbonate-silicate cycle acts as an effective buffer against changes in the global temperature. On Venus, obviously, the buffer failed.

The reason it failed, quite simply, is that the carbonate-silicate cycle requires the existence of oceans. But Venus lost most of its water early on through a process known as the Jeans escape mechanism. Water molecules in the atmosphere are broken apart by ultraviolet radiation and form fragments known as free radicals, which in turn recombine to form molecular hydrogen and oxygen. The hydrogen, because of its light weight, quickly seeps away into space. Only the oxygen is retained, and this combines with surface rocks. What is important on the balance sheet is the net loss of water.

At low altitudes the Jeans escape mechanism is inefficient because of drag from other molecules. For all practical purposes, then, the escape mechanism is a high-altitude phenomenon. On the Earth, most water vapor condenses out in the "cold trap" between 9 and 17 kilometers. This means that there is very little water present at high altitudes to be lost—indeed, the stratosphere is bone dry, containing only 0.0004 percent of the Earth's water by volume. On primitive Venus, things were otherwise. Because of its greater proximity to the Sun, the planet receives twice as much solar radiation per unit area as the Earth. Thus conditions near the surface were hot and wet, and the planet's "cold trap" was pushed upward to a height of as much as 100 kilometers above the planet. Here the Jeans escape mechanism was able to proceed with sinister efficiency, robbing the planet of all of its oceans within a few hundred million years of its formation.

With no oceans, the process of calcium carbonate formation came to an end. (We do not know whether there ever were marine organisms on Venus, but even in their absence, direct precipitation of calcium carbonate crystals would still take place.) Thus there was nothing to check the buildup of carbon dioxide in the atmosphere. At present, the planet's atmosphere contains all of the carbon dioxide; none of it is locked up in rocks, as on the Earth. Thus Venus has a crushing surface pressure ninety times the Earth's—equal to that at a depth of some 1,200 meters below the surface of the ocean.

The Venusian clouds, which are made up of sulfuric acid droplets, are another consequence of the planet's early water depletion. These brilliant clouds are, in fact, the planet's shroud of death. Sulfur gases are readily dissolved in water, but once Venus lost its oceans they were able to remain at large. Indeed, the fact that they are still present means that the rate at which they are being produced through volcanic processes on the surface (see below) must exceed the rate at which they are being removed by combination with calcium minerals in rocks.

Within a few years of the radar disclosure of the Venusian rotation, radio astronomers began producing the first maps of the surface itself. Most of the planet was shown to consist of rolling plains, but several elevated "continental" regions were also revealed. The first maps were only rough impressions, but much better results were subsequently obtained by radar-equipped orbiting spacecraft, including the American Pioneer Venus, which went into orbit around Venus in December 1978, and the Soviet Veneras 15 and 16, which arrived in October 1983 and which obtained detailed radar imagery of the northern hemisphere of the planet. By far the most detailed mapping has been done by the Magellan spacecraft (see fig. 6.4), whose resolution is 120 meters, which is ten times better than that of the Venera spacecraft. Magellan was launched from a Space Shuttle in April 1989, and on August 10, 1990, entered orbit around Venus. The orbit was highly elliptical, carrying the spacecraft to between 294 and 8,450 kilometers of the surface, and was inclined to the equator by an angle of 85°. The plan was to scan the planet in a series of strips—each measuring 20–25 kilometers wide by 15,000 kilometers long—by bouncing radio waves off the surface for a period of about half an hour each time the spacecraft was at the closest point of its orbit to Venus. The first strip maps, of the southern hemisphere plain known as Lavinia Planitia, were received in September 1990, and by mid-May 1991, after 1,789 orbits of the planet, 84 percent of the planet's surface had been mapped.

The surface revealed by this mapping project is mostly volcanic. Vast

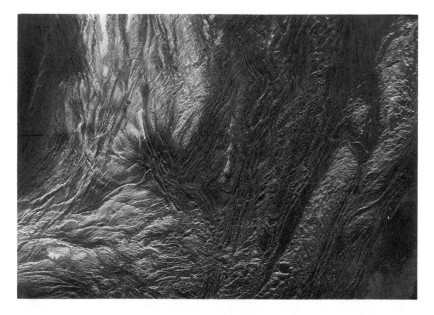

FIGURE 6.4. A radar image of Venus obtained by the Magellan spacecraft. The region shown here covers the Freyja Mountains and, at the bottom, the northernmost part of the volcanic plain Lakshmi Planum. (NASA photograph)

lava-flooded plains, volcanic plateaus, large shield volcanoes, and tens of thousands of smaller volcanic domes are scattered across the surface. However, contrary to expectations, there are no features formed by plate tectonics, as on the Earth. Venus has a style of its own in which it seems that the hot, churning mantle catches the thin overlying crust and drags it about, here piling it into plateaus and mountains, there dividing it into great rifts, of which Devana Chasma and Diana Chasma are spectacular examples.

In the northern hemisphere the most interesting region is the continental area known as Ishtar Terra. It is located at 60°N latitude, and it covers an area about equal in size to the continental United States. Much of Ishtar Terra consists of a volcanic plateau, Lakshmi Planum, which measures 2,500 kilometers across and which is capped by two large shield volcanoes, Colette and Sacajawea. Towering mountain ranges surround the plateau on every side; to the east rise the Maxwell Montes, to the north and northeast the Freyja Montes, to the west the Akna Montes, and to the south the Danu Montes.

The greatest heights are attained by the peaks of the Maxwell Montes.

(This is, incidentally, the only feature on Venus named for a man—the Scottish physicist James Clerk Maxwell. It was named before 1971, when the convention was adopted of calling all the main features after noted women of mythology and history.) Some of the peaks of the Maxwell Montes reach as much as 9 kilometers above the level of Lakshmi Planum, or 12 kilometers above Venusian "sea level," so they are much higher than the Himalayas of the Earth and no less rugged.

Just east of the Maxwell Montes is the interesting 100-kilometer-wide structure Cleopatra. It was formerly thought to be a volcanic caldera, but as Magellan has now shown, it is in fact a double-ringed impact crater. Still farther east the parallel ridges of the Maxwell Montes give way to a slightly elevated region of complex formation known as a tessera. This appears to be an older section of crust that has been deformed numerous times since its formation. Finally there are the so-called coronae, large ringlike structures of which about thirty were found in the plains around Ishtar Terra by Veneras 15 and 16. Each corona consists of an elevated and extensively fractured region surrounded by a ridge, and according to one view, each may lie over hot plumes in the planet's mantle where partially melted material has welled up and caused extension and fracturing of the overlying crust.

Also in the northern hemisphere, at latitude 30°, is an upland of a different type—Beta Regio. It is a domelike region some 2,000 kilometers across, and it is the site of two more large shield volcanoes, Rhea Mons and Theia Mons, as well as the great rift Devana Chasma, which has been partially flooded by lava flows from Theia Mons.

The southern hemisphere is dominated by the largest of the "continental" areas, Aphrodite Terra. Located at latitude 30°s, it is about equal in size to Africa. It is marked by several enormous rifts, of which Diana Chasma is on the scale of the Valles Marineris of Mars. At its greatest breadth it measures 280 kilometers across, and its bottom is 4 kilometers below the adjacent ridges.

Only about 10 percent of the planet's surface consists of high continental regions. Smooth, lava-flooded plains cover the rest, with here and there large shield volcanoes looming above them. From the number of impact craters found on these plains, their average age has been put at about 400 million years. However, there are large areas where there are no impact craters at all, indicating that they must have been flooded within the past few tens of millions of years, a blink of the eye on the geologic time scale.

There is, then, every reason to believe that the planet is still geologically active. The Magellan spacecraft has yet to find direct evidence of volcanic activity, but there is plenty of indirect evidence—though because the rate at which new crust is being produced appears to be only about a sixtieth of the Earth's, Venus is far less active than our own planet. (An active volcano on Venus, by the way, is not likely to produce an explosive eruption like Vesuvius, Krakatoa, or Pinatubo on the Earth. Because of the enormous atmospheric pressure, it would simply ooze lava gently onto the surface.)

Even the impact craters on Venus have their unique features. The largest such crater so far identified is a double-ringed structure 275 kilometers across, for which Magellan scientists have proposed the name Mead. But there are no craters at all smaller than about 3 kilometers, for the simple reason that small meteorites are unable to penetrate the dense atmosphere. The ejecta blankets of many impact craters have rounded lobes, which gives them the appearance of splash patterns—indeed, debris ejected from impacts behaves like a liquid on passing through the thick Venusian air. But the outstanding feature of the impact craters on Venus is the remarkably pristine appearance of most of them. This may be somewhat surprising, given the harsh conditions that reign at the surface. However, as there is scarcely any surface wind and no water on Venus, its features are actually subject to very little erosion in the usual sense. Moreover, given the dense atmosphere, these features might as well be buried at the bottom of the sea. It seems, then, that the impact craters on Venus are erased mainly through flooding by lava flows.

Twenty-three spacecraft have been launched toward Venus, beginning with the unsuccessful Venera 1 of 1961 and the successful Mariner 2 a year later, which first established the inhospitable conditions on the surface. Of this large retinue of spacecraft—more than for any other body except the Moon—fifteen have actually attempted landings on the hellish surface.

During the 1960s, Veneras 3, 4, 5, and 6 were all crushed during their descent through the atmosphere. The first to arrive on the surface intact was Venera 7, which transmitted data for twenty minutes in December 1970. It was followed in March 1972 by Venera 8, which also remained in radio contact briefly after landing in the plains just south of the equator.

The dramatic Venera 9 and 10 missions of October 1975 had both orbiter and lander components. The landers arrived at a pair of sites just east of Beta Regio, in the northern hemisphere, and remained in radio contact for about an hour. Moreover, each returned a television picture of the surrounding landscape. Floodlights had been provided for the purpose, but in the event there was enough light so that they did not have to be used. Venera 9 photographed a boulder-strewn scene, Venera 10 an area of flat stones. Winds at both sites were very light.

Veneras 11 and 12 landed on Venus in December 1978, but they were overshadowed by Veneras 13 and 14, which in March 1982 landed in Nava Planitia east of Phoebe Regio, near the equator, and returned the first color pictures from the surface. The sky appeared an ominous dull orange. The rocks, after due allowance was made for the orange sky cast, proved to be dark and colorless. A careful spectral analysis showed that they resembled oxidized basalts and would have appeared rust red on the Earth. At 500°C, even rust-red rocks appear dark and colorless. Additional probes were dropped to the surface in June 1985 as part of the Vega 1 and 2 combined mission to Venus and Halley's Comet. While some of the equipment failed on Vega 1, the Vega 2 lander analyzed soil samples at a site just north of Aphrodite Terra. Again the results indicated a resemblance to the basalts of the Earth's oceans, and indeed by now the volcanic origin of the planet's surface materials is very well established.

To date, the lone American probe to reach the surface intact was one of four that plummeted through the atmosphere of Venus as part of the Pioneer Venus mission, of which the radar mapper was the other part. It unexpectedly continued to transmit for more than an hour after settling on the surface on December 9, 1978.

The probes will lie forever beneath the unrelenting clouds, lit by the dim orange glow of the Sun. Never will the stars shine on them again. The rolling banks of sulfuric acid clouds, hanging above them as impervious as a ceiling, keep any view of the stars from reaching the surface. If there were any beings there, they might never realize anything about the larger universe beyond.

One might wish, in the words of Dante in the *Inferno,* "only to look, and pass on" for we have seen Venus now for what it is. Though it looks so alluring across the distance, it is a desolate, forsaken world, and our own Earth is in reality far more lovely. But Venus tells us a highly cautionary tale. Greenhouse gases have been building up steadily on the Earth, the byproducts of human activity. Though there is still debate over whether greenhouse warming has already been detected—on balance the evidence seems to suggest that it has—there can be no doubting the principle, and it is only too probable that the stage has been set for a significant degree of global warming well into the next century.

This may be so, but in the long run the greenhouse effect produced by human activity will be only a temporary aberration. Fossil fuels are, after all, not unlimited, and once they run out, the carbonate-silicate cycle must inevitably push things back in the direction of equilibrium again. The Earth will thus survive man, though man may not survive man. But with or without human tampering, the Earth is not immortal; its eventual demise is certain. The Sun, as we know, continues to increase its luminosity by about one percent every hundred million years, and as it does so, the Earth will find it more and more difficult to hang onto its precious supply of water. It is now the jewel of the planets; in another billion years it may become another Venus. What the Earth now is, Venus, in the days of its long-vanished oceans, may have been; what Venus is, the Earth will be—the sobering moral of planetary evolution.

CHAPTER SEVEN

Mars

Mars, because of its intense reddish color, was called Nirgal (Star of Death) by the Babylonians, while the ancient Greeks and Romans named it for their god of war. It has played an especially important role in the history of human ideas. Notably, from Tycho's observations of Mars, Kepler first discovered the elliptical shape of the planets' orbits. As it describes its ellipse, Mars approaches to within 206,615,000 kilometers of the Sun at perihelion and then recedes to 248,990,000 kilometers at aphelion. The planet completes each orbit in 687 Earth days, 669 of its own.

Mars is best observed around the times of opposition, when the minimum separation between the Earth and Mars takes place. Oppositions occur at intervals of about 2 years, 2 months, but not all of them are equally favorable. Because Mars's orbit is more eccentric than the Earth's, it is mainly Mars's position in its orbit that determines the opposition distance, which varies from 55.7 million kilometers if Mars is near perihelion to 101.3 million kilometers if it is near aphelion (fig. 7.1). The most favorable oppositions occur only once every fifteen or seventeen years. The last time was in 1988, the next will be in 2003.

The first observations of surface features on Mars were made in 1659 by Christiaan Huygens of Holland, who sketched the dark area that became known on later maps as Syrtis Major. Huygens set the rotation at about 24 hours, which G. D. Cassini later corrected to 24 hours, 40 minutes. The currently accepted figure is 24 hours, 37 minutes, 22.6 seconds.

Huygens and Cassini also recognized the planet's bright polar caps, and by the end of the eighteenth century William Herschel had shown that the

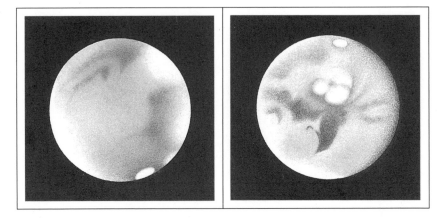

FIGURE 7.1. Two views of Mars by Richard Baum. The drawing at left was made during the planet's unfavorable opposition of 1980. The north polar cap is displayed, and the dark region extending southward from it is the Mare Acidalium. The drawing at right, made during the very favorable opposition of 1988, shows an extraordinary amount of detail. The prominent dark feature at the center is Syrtis Major, and the cloverleaf-shaped bright area above it is Hellas, an old impact basin. The dark spot at the center of Hellas was once thought to be a lake and was given the name Zea Lacus by Schiaparelli. The south polar cap is also shown. (Drawings based on observations made at Chester, England, using a 4 ½-inch refractor at 186×. Left: March 13, 1980, 19:35 UT; right: Oct. 20, 1988, 17:20–18:30 UT)

obliquity of the Martian axis, or the angle it makes from the perpendicular to the orbit, is almost identical to the Earth's. Thus the Martian seasons are analogous to ours, though more extreme in the southern hemisphere than in the northern, as the southern hemisphere is tipped toward the Sun when the planet is near perihelion and away from the Sun when it is near aphelion. Accordingly the southern hemisphere has long, bitterly cold winters and short, hot summers. Herschel also observed clouds in the planet's atmosphere and concluded that "the inhabitants . . . probably enjoy a situation similar to our own."

At the beginning of the nineteenth century, Schroeter's observations convinced him that the various light and dark areas on the planet were the changing forms of a mere "shell of cloud," but as the century grew older it became evident that there were stable surface features as well as clouds. To some, the outlines of the surface features were more than vaguely reminiscent of the Earth's oceans and lands. The Jesuit astronomer Father Pierre Angelo Secchi, for instance, wrote: "Everything is variegated like a map of the Earth." Thus the analogy to the Earth became even more compelling. This set the stage for the favorable opposition of 1877, a landmark year in the study of the planet.

The first development was the discovery of two tiny satellites by Asaph Hall using the recently unveiled 26-inch Clark refractor at the U.S. Naval Observatory in Washington, D.C. Unsuccessful searches for satellites had already been mounted by Herschel and Heinrich d'Arrest, both highly skilled observers, and at first Hall, too, met only with discouragement. He was at the point of abandoning the search when his wife, Angelina, urged him to return to the dome for one last try. He describes what happened next:

> I began to examine the region close to the planet, and within the glare of the light surrounding it. This was done by keeping the planet just outside the field of view, and turning the eyepiece so as to pass completely around the planet. While making this examination on the night of August 11, I found a faint object on the following side and a little north of the planet, but had barely time to secure an observation of its position when fog from the Potomac River stopped the work.

Hall had found the outer satellite, Deimos. In those days, the observatory was located in the lowlands along the Potomac, not yet having been moved to its present site in the northwest part of the city. Thus fog interfered with astronomical viewing much of the time, and not until August 16 did Hall have another suitable night. He then recovered his suspected satellite and confirmed that it was accompanying Mars in its motion. The next night he discovered Phobos, the inner moon. As it appeared on different sides of Mars during a single night, he at first supposed that there might be two or even three inner moons, but watching through an entire night settled the matter—there was, in fact, but one.

Phobos and Deimos are peculiar objects. Phobos lies at a distance of only 9,400 kilometers from the center of Mars, 6,000 from the surface itself. From Phobos, Mars would be an astounding sight. Its disk would subtend some 43° and would fill half the sky from the horizon to the zenith. Features down to a few kilometers in breadth would be readily visible with the naked eye, and the view of the great Martian volcanoes and canyons would be glorious.

Phobos is spiraling ever closer to Mars owing to the effects of tidal friction on its odd-shaped body (it is shaped rather like a potato and measures 27 by 19 km). Eventually it will fall into the planet, though not for at least forty million years. The present period of revolution is 7 hours, 39 minutes, so the moon completes three full revolutions in the time Mars takes to rotate once on its axis. As a result, it rises in the west and sets in the east, and it remains above the horizon for only about 4 1/2 hours at a time.

Because the orbital inclination of Phobos is only about one degree, for all practical purposes it lies in an equatorial orbit. It is frequently eclipsed by the planet's shadow, and observers on Mars beyond latitudes 70°N or S would be unable to catch sight of it at all, as it would never clear the horizon.

Deimos, slightly smaller than Phobos at 15 by 11 kilometers, lies at a distance of 23,500 kilometers from the center of Mars, and its orbit is also nearly in the equatorial plane. Its period of revolution is 30 hours, 18 minutes, so, unlike Phobos, which scurries across the sky, Deimos remains above the horizon for 60 hours at a time. From Mars, Phobos would appear about as bright as Venus as seen from the Earth, while Deimos would rival such bright stars as Arcturus or Vega.

Even in a large telescope, Phobos and Deimos appear as tiny specks of light. Phobos is especially difficult to see because it lies so close to the glare of the planet, but I found Deimos surprisingly easy to capture with my 12 1/2-inch reflector in 1988. Needless to say, only a few years ago it would have been hard to imagine that we would ever be able to map the surfaces of these dwarf moons in detail. By now, however, they have been examined at close range by Mariner 9, Viking Orbiter, and Phobos 2—the last, a Russian probe, fell silent in March 1989 just as it was nearing the satellite, though useful preliminary results were obtained.

The largest crater on Phobos is Stickney, which honors the maiden name of Angelina Hall. It is 10 kilometers in diameter, and the impact that formed it, which must have come very close to shattering the satellite, produced a series of peculiar ridges and groovelike strains in the surface. The main ridge is called Kepler Dorsum. After Stickney, the main craters on Phobos are Hall, Roche, Todd, Sharpless, and d'Arrest. On Deimos the largest impact craters, Swift and Voltaire, are 3 kilometers across.

There can be little doubt that Phobos and Deimos represent captured asteroids. It might be thought that their near-equatorial orbits argue against this, but they are so close to the planet that tidal effects would have damped their orbital inclination to their present values long ago.

Aside from the discovery of the Martian satellites, the other important development of the opposition of 1877 was the result of a careful study of

FIGURE 7.2. Giovanni Virginio Schiaparelli
(1835–1910) (Courtesy Yerkes Observatory)

the planet by G. V. Schiaparelli in Milan. With an 8.6-inch refractor, Schiaparelli set out to produce an accurate map of the Martian surface, and indeed he was the first to do so on the basis of careful measurements of the positions of various features. Moreover, he devised an innovative system of nomenclature.

Earlier maps had given to Martian features the names of astronomers. For example, on that drawn up by England's Richard A. Proctor in 1867, there had been a Cassini Land, a Mädler Continent, and a Beer Sea, and no less than six features bore the name of W. R. Dawes, a noted English observer of the day. Overall, in fact, Proctor's map was rather chauvinistic, and as a result it failed to win much favor on the Continent. Schiaparelli, an armchair classicist who wrote Latin verses for pleasure, proposed instead a system that drew on mythology and the geography of the classical world for its names. Solis Lacus, the Lake of the Sun, recalled Homer's Bath of the Ocean, from which the Sun rose each morning. Nearby was Aurorae Sinus, the Bay of Dawn, followed by Margaritifer Sinus (the rich Pearl-Bearing Gulf of the Indian coast), Syrtis Major (the Gulf of Sidra), and Mare Tyrrhenum (the Tyrrhenian Sea, west of Italy). The bright areas included Ausonia (Italy) and Hellas (Greece). In the far west were Elysium, Mare Cimmerium, and Atlantis. The poetry of these

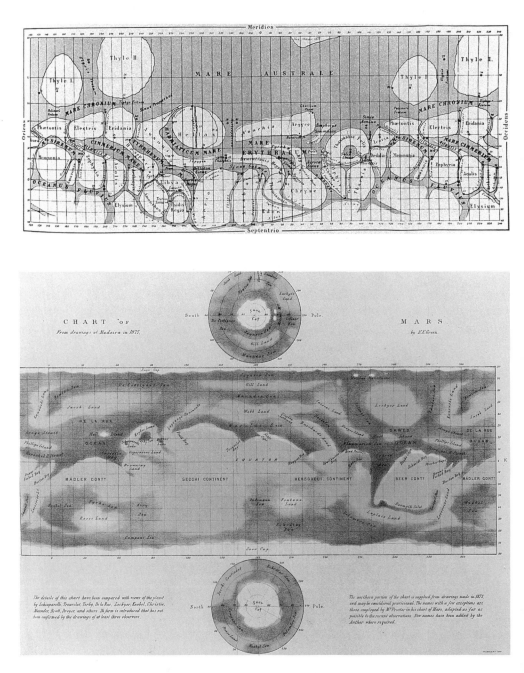

names could not be denied, and they soon came into general acceptance.

In addition to charting the main dark and light patches on the planet, which were then widely believed to be the seas and continents of Mars, Schiaparelli glimpsed a network of fine linear features on the surface, to which he gave the name *canali*. It has often been pointed out that in Italian *canali* means simply "channels," which implies nothing artificial, but it was as "canals" that it was translated into English, with the predictable result.

It is highly instructive to compare Schiaparelli's 1877–78 map with that made during the same opposition by the English observer Nathaniel E. Green, who used a 13-inch Newtonian reflector at Madeira (see fig. 7.3). I like to think of them as left- and right-brain views of Mars, respectively. Whereas Schiaparelli had originally been trained as an engineer and thus was adept in the use of "rule and compass," Green was a professional portrait artist who once gave lessons in painting to Queen Victoria. A glance at the two maps suggests where at least part of the blame for the ensuing Martian "canal" episode can be laid. As Sir Ernst Gombrich has written in his book *Art and Illusion,*

> The artist, clearly, can render only what his tool and his medium are capable of rendering. His technique restricts his freedom of choice. The features and relationships the pencil picks out will differ from those the brush can indicate. Sitting in front of his motif, pencil in hand, the artist will, therefore, look out for those aspects which can be rendered in lines—as we say in pardonable abbreviation, he will tend to see his motif in terms of lines, while, brush in hand, he sees it in terms of masses.

FIGURE 7.3. These maps of Mars show the impressions two skilled observers obtained during the opposition of 1877–78. The map at top is by Giovanni Schiaparelli, using the 8 ¼-inch refractor of the Brera Observatory in Milan. It records Schiaparelli's first impressions of the curious markings he termed *canali*. The map at bottom is by Nathaniel Green, an English astronomer who was observing with a 13-inch reflector on the island of Madeira in the North Atlantic. (From Camille Flammarion, *La Planète Mars, et ses conditions d'habitabilité,* vol. 1 [Paris, 1892]; and N. E. Green, "Observations of Mars, at Madeira in Aug. and Sept., 1877," *Memoirs of the Royal Astronomical Society* 44 (1877–79), fol. p. 138)

Anyone who has ever studied Mars with a telescope will probably agree that Green comes closer to representing the planet's appearance accurately, but Schiaparelli's schematic had far greater psychological impact, and during the 1890s, "canals" became all the rage.

Schiaparelli's later maps showed the "canals" as ever more regular and geometric, and he also described the strange phenomenon he called gemination: sometimes a single "canal" would appear in a given course, and at other times there would be two running side by side in parallel.

At first Schiaparelli himself hewed firmly to the view that the "canals" were natural surface features, and even as late as 1893 he was still writing: "The network formed by them was probably determined in its origin in the geological state of the planet. . . . It is not necessary to suppose them the work of intelligent beings." Two years later, however, he conceded that the geminations in particular were so inexplicable that the idea that the activity of intelligent beings was behind them "ought not to be regarded as an absurdity."

By then the center of Martian activities had shifted away from Milan. Schiaparelli's eyesight was no longer as good as it had been, and Camille Flammarion's observatory at Juvisy, south of Paris, and Percival Lowell's at Flagstaff, Arizona, had been set up in the meantime. While Schiaparelli ventured into speculation only on rare occasions, Flammarion and Lowell founded their observatories specifically with the idea of investigating the possibility of life on Mars—and they certainly made no attempt to conceal their fascination with the idea. Just as the remarkable Martian canal hysteria of the 1890s was getting underway, Flammarion wrote:

> It seems that the continents of Mars may be flatter than ours, and formed nearly everywhere by vast plains; for, on the one hand, the seas often run to immense distances and contract in the same proportion and, on the other hand, the straight lines or canals . . . show us a sort of geometrical network extending all over the continents. . . .
>
> What a strange geography! The future will doubtless throw light on this mystery.

The idea that Mars is much flatter than the Earth was, by the way, widely accepted then and later, and it was a necessary assumption underlying some of the radical theories about the planet that were soon being proposed. The ebullient French astronomer asked, somewhat rhetorically: "Could the meteorological circulation which produces on this planet, as on ours, seasons, fogs, snows, rain, heat and humidity . . . act for thousands of centuries without giving birth to the smallest blade of grass?" Calling attention to the predominance of yellows and oranges on Mars, he further invited readers to "enlarge the circle of botanical conception, and admit that vegetation is not necessarily green in every world."

Even more compelling as a spokesman for the idea of Martian life was Percival Lowell. A Boston blue-blood, he became a millionaire by the age of thirty by managing his grandfather's textile business, then abruptly changed course and spent the next ten years traveling in the Far East. By the time he returned home from his last trip there in late 1893, he had developed an all-consuming passion for the red planet.

With his imagination captivated by Schiaparelli's discoveries, Lowell decided to carry out his own study of Mars, but there was no time to waste—the last really favorable Martian opposition of the century was due in October 1894. He had strong views about where observatories ought to be located, and tests showed that conditions in Flagstaff were likely to be favorable, so off to Flagstaff he went, bringing with him a borrowed 18-inch refractor. At the end of May 1894 he first threw open the shutters of his dome and swung his telescope toward Mars.

An examination of Lowell's observing notebooks shows that he came to Flagstaff sharing the preconception of most astronomers of his day that the dark areas on Mars were seas. This was only natural, given that under the best seeing conditions they appeared bluish green. However, when Lowell's assistant, William H. Pickering, trained an instrument called a polariscope on Mars, he found that the light reflected from the dark areas was not polarized, as it ought to have been if it were reflected from water. This meant that the bluish green areas could not be seas. So much was clear, yet how then to account for the definite changes that had been observed in their outlines, which hitherto had been interpreted as due to watery inundations or retreats?

In his 1895 book *Mars* and in other writings, Lowell vividly described the changes he himself had observed. In *Mars* he wrote:

For some time the dark areas continued largely unchanged in appearance; that is, during the earlier and most extensive melting of the snow-cap. After this their history became one long chronicle of fading out. Their lighter parts grew lighter, and their darker ones less dark. For, to start with, they were made up of many tints; various shades of blue-green interspersed with tints of yellow. . . .

The long, dark streaks that in June had joined the Syrtis Major to the polar sea had by October nearly disappeared; in their southern parts they had vanished completely, and they had very much faded in their northern ones. . . .

It will be seen that the extent of these changes was enormous. . . . A wholesale transformation of the blue-green regions into orange-ochre ones was in progress upon that other world.

Lowell could think of only one way of explaining these observations: the bluish-green areas were tracts of vegetation, having their own cycle of growth and decay. As for the yellowish areas, they became in his mind deserts lonely and desolate almost beyond imagining. Clearly the Arizona deserts south of the observatory had made an impression on him:

> Beautiful as the opaline tints of the planet look, down the far vista of the telescope-tube, they represent a really terrible reality. To the bodily eye, the aspect of the disk is lovely beyond compare; but to the mind's eye, its import is horrible. . . . All deserts, seen from a safe distance, have something of this charm of tint. Their bare rock gives them color, from yellow marl through ruddy sandstone to blue slate. And color shows across space for the massing due to great extent. But this very color, unchanging in its hue, means the extinction of life. Pitilessly persistent, the opal here bears out its attributed sinister intent.

The planet, Lowell argued in his appealingly romantic prose, was well on its way to drying up. The only sources of precious water were the snowcaps at the poles, whose thinness was proved by the rapidity of their melting, and which gave rise each spring to a narrow blue band that Lowell identified as meltwater. Given the dearth of water on Mars, Lowell suggested that if beings of sufficient intelligence inhabited it, they would be forced to resort to a system of irrigation—an engineering project easier to carry out on Mars than on Earth because of the supposed flatness of the planet. Such an irrigation system, he stressed, would presumably look precisely like that which Schiaparelli had already identified there, and Lowell himself proceeded to document it in ever greater detail.

Not all astronomers agreed with Lowell's description of Mars, and during the fine opposition of 1909 the debate over the "canals" reached something of a climax. Lowell, interestingly enough, almost always stopped down the 24-inch refractor he used from 1896 onward (though in his defense it must be admitted that stopping down can improve the image if atmospheric seeing conditions are not first-rate), and he so advised E. M. Antoniadi, Flammarion's onetime assistant at Juvisy, who was given permission to use the 32 3/4-inch refractor at Meudon, near Paris, in 1909. Antoniadi had strong views of his own, however, and he ignored Lowell's advice. On September 20, 1909, his first night with the great refractor, Antoniadi enjoyed several hours of "glorious seeing" in which he found that the Martian deserts resolved into "winding, knotted, irregular bands, jagged edges of half-tones, or other complex details"—almost everything, that is, except canals. Antoniadi's observations did not provide the final coup de grace to the "canals"—Lowell himself remained firmly committed to them right up to his death in 1916—but they certainly weakened the case for them.

We now know that there never were any true canals; they were an illusion due to perceptual effects by which complex details are seen under symbolic forms. Some of these complex details are actually streaky, as images from charged couple devices obtained in 1988 and 1990 show—though they look entirely natural, I hasten to add.

Canals or no canals, the Mars most astronomers accepted up until the mid 1960s was in many respects essentially Lowellian in inspiration (even Antoniadi was thoroughly convinced of the vegetation theory). Specifically, most astronomers would have accepted the following points:

1. that the polar caps were frozen water;
2. that there was an atmosphere whose pressure was probably right around 100 millibars, compared with 1,013 millibars on Earth;
3. that the daytime temperature, though colder than the Earth's, reached above the freezing point of water except in winter and in the extreme polar regions.

In a word, although the Martian conditions seemed harsh, they were considered far from totally inhospitable. Above all, there were the bluish green areas, which behaved, as Lowell had pointed out, just as areas of vegetation should. Their seasonal changes, which included an apparent "wave of darkening" spreading toward the equator with the melting of the polar cap in spring, seemed hardly explicable otherwise. Higher forms of life were not expected, but primitive organisms—perhaps something on the order of terrestrial lichens—seemed only too likely.

Over the years, an occasional voice of dissent was heard. There was the nineteenth-century Irish physicist G. Johnstone Stoney, for example, who suggested that the polar caps were not snow but instead frozen carbon dioxide—dry ice. But the blue "melt band" hugging the retreating polar cap seemed to disprove this, inasmuch as carbon dioxide, under the low atmospheric pressures of Mars, would not melt but would instead pass directly from the solid to the gaseous state through a process known as sublimation. This being the case, water and water alone seemed capable of accounting for the observations.

Another unorthodox but prescient scenario was painted by Dean B. McLaughlin of the University of Michigan. In the 1950s McLaughlin pointed up the dissimilarity between the forms of the Martian dark areas and the circular basins on the Moon. "The sharpish ends of features such as Syrtis Major," he wrote, "point into the wind, and are essentially point sources of some dark material that is carried from these points, fanning out because of variable wind direction. If we restrict ourselves to natural phenomena of which we have experience on Earth, the point sources can have but a single interpretation: they are volcanoes whose ash is carried by the winds and deposited in the pattern we see." This was an inspired suggestion and proved to be right on many counts. Its main flaw was its assumption of active volcanism on Mars. In any case, McLaughlin's ideas received little support at the time. Instead, the Lowell-inspired version of Mars, with its tracts of vegetation and lonely deserts, remained in vogue right up until July 1965.

In that month the Mariner 4 spacecraft swept past the planet. What it showed came as a complete shock. The spacecraft's stark black-and-white images revealed a desolate world scarred with craters, resembling the Moon in every respect. Moreover, it was clear from Mariner's measurements that the thickness of the Martian atmosphere had been vastly overestimated: in reality, the surface pressure was on the order of a mere 5 to 10 millibars. Two more spacecraft, Mariners 6 and 7, passed by the planet in 1969 and did nothing to change this dour impression.

True, the three flyby spacecraft had photographed only a tiny fraction of the Martian surface, and it was, of course, quite possible that the parts of Mars they had sampled were not representative of the whole planet. In fact, this proved to be the case, as Mariner 9 showed during its extended mission to Mars in 1971–72. When Mariner 9 went into orbit around Mars in November 1971, it found the whole planet in the throes of a great dust storm. Only the south polar cap and four mysterious dusky spots were visible through the cover.

Clouds on Mars are of two main types: yellow dust clouds and whitish clouds made up of condensates of water or carbon dioxide. For the moment we will say no more about the latter, but when Mars is near perihelion, yellow dust clouds tend to form over elevated plateaus (for example, Hellespontus, Noachis, and Solis Lacus), and over a few weeks they may give rise to planetwide storms. The great 1971 storm, for example, began in mid-September with a yellowish streak over Helle-spontus, but by early October the familiar Martian surface features were completely veiled, and not until late December did the dust finally begin to clear.

The evolution of the great dust storms is of considerable interest in its own right, and I shall have more to say about this later; but from the point of view of Mariner 9, the dust storm that greeted it on its arrival was merely a nuisance. Indeed, the same storm completely spoiled the Soviet Mars 2 and 3 missions to the planet. Unlike their American counterpart, they had been preprogrammed, and thus when they arrived in Martian orbit just after Mariner 9, they could only carry out their commands pointlessly while Mars was covered with dust from pole to pole.

After the Martian atmosphere cleared, the four dusky spots of Mariner 9's early sequences were found to be towering shield volcanoes. The largest, Olympus Mons, had been glimpsed in the old days by Schiaparelli, who noted a whitish patch and named it Nix Olympica, the Snows of Olympus. It rises 25 kilometers above the surrounding plains and spans some 700 kilometers. The other three volcanoes, Ascraeus Mons, Pavonis Mons, and Arsia Mons, are also larger than the largest such feature on Earth, the Hawaiian shield volcano Mauna Loa. Whitish clouds had frequently been observed in this region, known as Tharsis on the classical maps. These proved to be orographic clouds, water-ice condensates that gather over high mountains. Another location over which such clouds tend to form is the elevated Elysium plateau to the west, the site of several smaller shield volcanoes.

Mariner 9 demolished the old notion of a flat Mars; there is obviously considerable relief there. Yet all is not mountainous that mountainous seems. Specifically, the Mountains of Mitchel, small detached areas left behind as the south polar cap retreats (they are named for Cincinnati astronomer Ormsby MacKnight Mitchel, who discovered them in 1846) proved to be not mountains but depressed areas where "snow" hidden from direct sunlight lingers.

Close to the Tharsis region and its towering shield volcanoes is a system of interconnected canyons, which begins just east of Arsia Mons and extends westward for some 3,800 kilometers—a full one-sixth of the circumference of the Martian globe. The classical maps showed a "canal" there, Agatho-daemon, but the modern name is Valles Marineris—the Valleys of Mariner (see fig. 7.4). Near Arsia Mons the canyons consist of short, deep gashes that intersect at all angles, making up the feature known as the Chandelier. To the east, the canyons become more contin-

FIGURE 7.4. A panorama of Mars created by Alfred McEwen at the
U.S. Geological Survey in Flagstaff, Arizona, from Viking Orbiter
spacecraft images. The view is dominated by the vast canyon system of
Valles Marineris, which extends for 4,000 kilometers, a sixth of the
way around the planet. The three large shield volcanoes at the extreme
left are, from bottom to top, cloud-filled Arsia Mons, Pavonis Mons,
and Ascraeus Mons. (Courtesy U.S. Geological Survey, Flagstaff,
Arizona)

uous and run in parallel as three main branches, with the combined width
across all three reaching 700 kilometers at one point. The southernmost
component joins up finally with the great Coprates Chasma. Still farther
to the east the canyons give way to a jumbled and broken surface known
as chaotic terrain.

These Martian canyons are not really analogous to terrestrial features
like the Grand Canyon of northern Arizona, which was cut by action of
the Colorado River. The Martian canyons were formed by collapse along
a fault line, no doubt in association with the Tharsis uplift to the west, and
thus resemble such features as Diana Chasma on Venus.

On the other hand, there are features on Mars that were formed by running water. The chaotic terrain itself appears to represent a wide area of collapse related to the melting of subsurface ice, and several large outflow channels originate here. Unlike the rilles on the Moon, which are collapsed lava tubes, these Martian channels really were cut by the action of running water and are the result of flooding on a catastrophic scale. The largest are the Ares, Simud, and Tiu Valles. These channels lack tributaries and can be compared with a different type of Martian channel, also cut by water, which is found especially in the heavily cratered areas. This type of channel begins with a few short, deep tributaries, runs typically for a few tens of kilometers, then abruptly stops. Most are rather small, but a few—notably the Ma'adim Vallis, 600 kilometers long, and Al Quahira, 800 kilometers—are imposing features.

The discovery of these water-carved channels was undoubtedly the most exciting discovery of Mariner 9, proving that Mars is not simply another Moon, though it has its heavily cratered areas too. There are, for example, several large impact basins. Hellas, at 1,800 kilometers across, is larger than the lunar Mare Imbrium or the Caloris basin of Mercury, and Isidis and Argyre are other examples of the type. Other large impact features worthy of note are the craters Schiaparelli, Huygens, Cassini, and Antoniadi, which range from 380 to 500 kilometers in diameter. Lowell, at 190 kilometers, is a fine example of a double-ringed crater with a cluster of prominent central peaks.

For the most part, the primitive, heavily cratered terrain is located in the Martian southern hemisphere, and it was over this that the flyby Mariners passed. They had been directed there because most of the principal dark areas of the planet are located there, and it had been on the dark areas that observers had chiefly focussed ever since Christiaan Huygens first sketched the Syrtis Major in 1659.

This points up a real irony, for we now know that this focus on the dark areas was somewhat misplaced. Though in a few cases, such as the Hellas and Argyre basins, differences in the surface reflectivity, or albedo, are topographically significant, for the most part they are not. Thus though such classical dark areas as Maria Cimmerium and Tyrrhenum, as well as Syrtis Major, Sinus Sabaeus, and Margaritifer Sinus, all lie in heavily cratered highlands, this is not true of Mare Acidalium, which occupies a low-lying level plain. The same inconsistencies are found in the bright areas. The region that contains the Tharsis volcanoes is the most youthful on the planet, but the bright expanse west of Syrtis Major contains some of the very oldest impact features.

The albedo differences on Mars seem to depend merely on whether the dusty covering is made up predominantly of iron oxides or of iron silicates. Moreover, in color the whole planet is a mottling of more or less intense shades of red; the dark areas are simply less vividly red than the bright areas. Observers saw shades of bluish green because of their visual physiology: a relatively neutral-toned area surrounded by an extended reddish field appears to the eye to be tinged with blue-green, an effect known as simultaneous contrast.

Not only are the dark areas reddish rather than bluish green, but the famed "wave of darkening" reported by earlier observers has now been shown to be, in fact, a wave of brightening. The bright areas in the southern hemisphere become brighter in the spring as fresh dust deposited during winter frost accumulation is uncovered by seasonal winds. The apparent darkening of the dark areas is, then, only a contrast effect. The visual observers were doubly deceived.

This is a good place to return to the Martian dust storms, which are intimately connected with the behavior of the polar caps. Because the planet receives 40 percent more radiation from the Sun at perihelion than at aphelion, the caps in the two hemispheres behave very differently. As noted earlier, the southern hemisphere has long, cold winters and short, hot summers. At its maximum extent, the southern cap covers a very large area. It is laid down as carbon dioxide in the atmosphere freezes out during the winter, and because the atmosphere is then crystal clear, the "snow" is pure white in color. During the hot summers, the carbon-dioxide snow rapidly sublimates away, but at the same time strong winds are whipped up wherever large temperature gradients are produced—for example, along the border of the cap itself, or in areas of high relief such as Hellespontus and Solis Lacus. These winds sweep dust into the atmosphere; the dust produces areas of cooling, which leads to more wind and the stirring up of still more dust. Thus a positive feedback loop is created that may eventually lead to a global storm. Interestingly, despite the rapidity with which the southern cap dwindles, it seldom disappears altogether. When the dusty season arrives, a small remnant of carbon-dioxide snow is usually protected from the blazing sun beneath the pall of dust.

The situation in the northern hemisphere is markedly different. Winter there occurs during the dusty season, and thus its carbon dioxide snowcap is much dirtier than the south's. This dirtier snow is a more efficient absorber of solar radiation. Moreover, summer comes to the northern hemisphere during the atmosphere's clear season, so it lies exposed to the direct solar rays. As a result, the carbon dioxide snow sublimates away entirely, and only a tiny kernel of water ice is left behind. It survives only because water ice melts at a temperature much higher than that at which carbon dioxide sublimates.

Since the Martian dust storms occur during the southern hemisphere's summers, that is, when Mars is near perihelion, they have been observed mainly during the years of the planet's most favorable oppositions, in 1909, 1911, 1924, 1941, 1943, 1956, 1971, and 1973. Some of my own observations from the 1988 opposition may be of interest. There were a few clouds in late June, but the atmosphere cleared up and remained clear from July to December. Indeed, it was so clear that even in a blue filter the dark areas were conspicuous (usually they are hidden in short wavelengths by haze). In early December, dust clouds again began stirring in the Martian southern hemisphere, but by then Mars was too far past perihelion—and too cold—to support a planetwide storm, and by mid December the atmosphere had cleared once again. In 1990, no major dust storm activity was observed.

Early cratering counts of the dry riverbeds on Mars suggested that they were relatively recent features, and it was even suggested that Martian conditions in the not very remote past may have been mild enough to have permitted water to flow on the surface. Subsequently, better cratering counts were made, and the new figures made it clear that the dry riverbeds are very old—older, at least, than the lava plains associated with the extinct Tharsis volcanoes. The last of them dried up perhaps 3.9 billion years ago, before the period of the heaviest meteorite bombardment was over. The dry riverbeds do prove, though, that during the first billion years or so of its history Mars must have had a much denser atmosphere than it does now. Because of warming by the greenhouse effect, its mean global temperature was then above 0°C, compared with the present −53°.

Recent calculations have shown that Mars may well have lost much of its atmosphere catastrophically inasmuch as a large asteroidal impact is more than capable of blowing away an entire planetary atmosphere. Alternatively, its atmosphere may have been lost through more gradual processes. With a diameter of only 6,787 kilometers, Mars is only half the size of the Earth. The smaller the world, the higher its surface-to-volume ratio and the faster it radiates its internal heat away into space. A consequence of this cooling is that a planet becomes less active geologically. True, in the Tharsis region the volcanoes, especially Olympus Mons, have continued building up through widely spaced eruptions over billions of years (their exact ages are uncertain). This region is unique, however.

Elsewhere on the planet, volcanism declined at a much earlier period, probably within the first billion years or so of the planet's life.

With this decline, Mars became less efficient at recycling carbon dioxide back into its atmosphere. Any carbon dioxide that was trapped in rocks through weathering remained trapped, and too little remained at large to produce a greenhouse effect strong enough to counteract the planet's greater distance from the Sun. Thus, according to this scenario, as the atmosphere disappeared, the planet froze over. In the case of Venus, it seems that its proximity to the Sun was the main factor that led to its death as a world, while Mars may have perished less because of its distance from the Sun than because of its small size.

Despite its desiccated and generally inhospitable condition now, the fact remains that Mars did have a dense atmosphere and running water on its surface during its first billion years, and this brings us to the all-important question, Did life develop there, and if so, is there a possibility that some especially hardy organisms may somehow have maintained a foothold?

Finding the answer to this question was the chief purpose of the two Viking missions of 1976. Each consisted of an orbiter and a lander component. The orbiters greatly extended the photographic coverage of the Martian surface and at a much higher resolution than Mariner 9 provided. Moreover, they did so for a period lasting, in the case of Viking 2, for over two Martian years (the spacecraft did not finally shut down until August 1980). Among the highlights were two global dust storms that enveloped the entire planet during 1977. But, naturally, interest was always greatest in what the landers would find on the surface itself.

The Viking 1 lander touched down at Chryse, the Land of Gold, on July 20, 1976, and Viking 2 at Utopia on September 3. The Chryse site was generally rolling, the Utopia site remarkably flat, and both were strewn with boulders believed to be debris thrown out from impacts. Though the specific impact feature responsible could not be identified in the case of Viking 1, Viking 2 lay within the ejecta blanket of the 90-kilometer-wide crater Mie to the east.

In addition to the photographs taken from the surface, the two Viking landers served as meteorological stations on Mars, sending back daily reports of Martian temperatures and weather conditions. Conditions were, as expected, frigid. Soon after its arrival, Viking 1 recorded a low of $-86°C$ ($-122°F$), which occurred at dawn. The high, $-30°C$ ($-22°F$), was recorded at 2 P.M. Local Martian Time. Since Viking 1's latitude was

22.5°N, these are Martian tropical conditions. Temperatures at the Viking 2 site, at 44°N, were on average some 5 to 10 degrees colder.

The atmospheric pressure at the Viking 1 site varied from 6.6 millibars in summer to 8.8 millibars in winter. The corresponding figures at Viking 2 were 7.4 and 10 millibars, reflecting its lower elevation. The variation in barometric pressure depends on the amount of carbon dioxide deposited in the polar caps. The south polar cap, in particular, because it fluctuates in response to its more extreme seasons, provides the dominant influence on pressures planetwide.

The biological laboratories aboard the two landers performed three experiments, but the results merely attested to the peculiar chemistry of the Martian soil, which is composed of long-irradiated iron and is rich in superoxides—thus its rust-brown color. No evidence of organisms was found, and the results of the mass spectrograph in particular, which failed to find any organic compounds despite having a sensitivity of one part in a billion, must be regarded as conclusive.

By the time the Viking landers came to rest on the planet, the search for life on Mars had admittedly become something of a will-o'-the-wisp. It was more a tribute to Percival Lowell and the hold his ideas about the planet still had over the human imagination than a sober assessment of probabilities. The remarkable similarities between the Earth and Mars that earlier observers thought existed have proved to be illusory: Mars is not another Earth, nor ever was. We are only now awakening from our long dream of Mars to see it as it really is. Still, with its great volcanoes and vast canyons, it remains, even if lifeless, a fascinating world.

CHAPTER EIGHT

Asteroids

In 1772 a German mathematician, J. D. Titius, found that if one takes the series 0, 3, 6, 12, 24, and 48, and adds 4 to each term, the resulting set of numbers closely approximates the relative distances of the planets from the Sun. Titius's discovery could hardly have been announced in less dramatic fashion; it appeared in a footnote to a book he was translating, Charles Bonnet's *Contemplation de la Natur*. But there it came to the attention of Johann Elert Bode, a well-known German astronomer of the day, and Bode immediately seized on its importance and did so much to popularize it that it eventually became known as Bode's law.

When Uranus was discovered by William Herschel in 1781, it too conformed to the scheme, and the "law" seemed to be established. However, there was something amiss: No known planet existed between Mars and Jupiter, in the position corresponding to the number 28. A few years after the discovery of Uranus, a Hungarian, Baron Franz Xaver von Zach, court astronomer to Duke Ernst of Saxe-Gotha and director of the Seeburg Observatory near Gotha, went so far as to attempt to compute an orbit for the missing planet. The only parameter that was unknown to him was, alas, the most important—its longitude, or actual location in the sky—and von Zach quipped: "I am having much the same success as the alchemists in their search for gold; they had everything except the vital factor."

In 1787 von Zach began working on a "revision of the stars in the zodiac," hoping to come across the missing planet in the course of this work. However, he soon came to the conclusion that the project was much too vast for one man. The following year he convened a conference at Gotha that both Bode and the French astronomer J. J. Lalande attended. Lalande suggested mounting a systematic search for the missing planet, with the work of scanning the zodiac divided among several astronomers.

Nothing was done to follow up on the idea at the time, but von Zach made plans for another conference in 1800. By then Europe was at war and traveling was difficult, but on September 17 six astronomers assembled at Johann Hieronymus Schroeter's observatory at Lilienthal. They were, in addition to Schroeter and von Zach, Karl Harding, Heinrich Wilhelm Matthäus Olbers, Ferdinand Adolf von Ende, and Johann Gildemeister. The meeting resulted in the establishment of the Vereinigte Astronomische Gesellschaft (United Astronomical Society), a confederation of astronomers dedicated to undertaking a systematic search for the missing planet. Schroeter was named director, and von Zach agreed to serve as standing secretary. The plan was to divide the zodiac into twenty-four search zones, with each zone to be scrutinized down to stars of the ninth magnitude. Twenty-four astronomers would share in the work, and in addition to the original members, such notables as William Herschel of England, Giuseppi Piazzi of Palermo, Charles Messier and Pierre Méchain of Paris, Barnaba Oriani of Milan, Johann Bode of Berlin, and Johann Huth of Estonia were invited to participate.

This was the first time that anyone had actually set out to search for an unknown planet—after all, Herschel's discovery of Uranus had been completely unexpected—and predictably enough, the plan did not escape ridicule. At Jena the German philosopher Friedrich Hegel was even then preparing a treatise for publication, *Dissertatio Philosophica de Orbitis Planetarum,* which argued that the attempt to search for a planet in order to fill a gap in a numerical series was the height of folly.

As it turned out, the ink was hardly dry on Hegel's essay when something turned up. On January 1, 1801, the first day of the new century, Giuseppi Piazzi, abbot of the order of Theatites in Palermo and director of the observatory founded there by Pope Pius VII, came across an eighth-magnitude "star" in Taurus that was not in his catalog. On returning to the same point in the sky the following night, he found that the object had shifted its position by 4 minutes of arc in right ascension and 3 1/2 minutes in declination. Piazzi, of course, had been one of the astronomers the Astronomical Society had hoped to enlist in its search. It is not clear whether the request to join the search for the planet had reached him before January 1. It does seem, however, that Piazzi already had at least a general knowledge of the search. Initially he released only two observa-

FIGURE 8.1. The first meeting of the Vereinigte Astronomische Gesellschaft, held at J. H. Schroeter's observatory at Lilienthal in September 1800 for the purpose of organizing a search for the "missing" planet between Mars and Jupiter. This painting records the meeting of the astronomers, led by Schroeter (shown bowing in the center), and Adolf Friedrich, prince of Great Britain and Hanover, and his delegation. The prince happened to be visiting Lilienthal while the meeting was underway. Schroeter's largest telescope, with its 20-inch aperture and 27-foot focal length, is shown in the background. (Painting in the Heimatverein Lilienthal; photograph courtesy Dieter Gerdes)

tions, made on January 1 and 23, despite having made fourteen observations during that period and eight more before the object was lost in the solar glare on February 11. He did not release the additional observations until the following August. It seems, then, that Piazzi was well aware of the fierce competition to discover the planet and was concerned with securing his priority—thus his desire to make it more difficult for other astronomers to make an exact verification. Piazzi knew only too well the probable significance of what he had found. Though in a letter to Bode written soon after the discovery he disingenuously referred to the object as a comet, he told Oriani of his strong suspicion that it was nothing less than a new planet.

Be that as it may, Piazzi's object had been lost in the twilight before he had made enough observations to work out an orbit using the standard methods. In July, when the planet was expected to emerge from the solar glare, William Herschel and others mounted telescopic searches but without success. It seemed that the object might never be recovered.

Just then the right man for the moment appeared. For some time the problem of calculating orbits, "without any hypothetical assumption, from observations not embracing a great period of time," had been the goal of a twenty-four-year-old German mathematician, Karl Friedrich Gauss. Gauss later wrote in the preface to his classic work, *Theoria Motus Corporum Caelestium*:

> Just about this time the report of the new planet, discovered on the first day of January of that year with the telescope at Palermo, was the subject of universal conversation. . . . Nowhere in the annals of astronomy do we meet with so great an opportunity—and a greater one could hardly be imagined—for showing most strikingly the value of this problem than in this crisis . . . when all hopes of discovering in the heavens this planetary atom, among innumerable small stars after the lapse of nearly a year, rested solely upon a sufficiently approximate knowledge of its orbit to be based upon these very few observations.

Gauss did work out an orbit, though the method he used was not exactly the same as that which he later published in *Theoria Motus*, as he there admits: "The methods first employed have undergone so many and such great changes, that scarcely any trace of resemblance remains between the method in which the orbit . . . was first computed, and the form given in this work." But by whatever means he arrived at it, Gauss calculated an orbit accurate enough for von Zach to recover Piazzi's object within half a degree of its predicted location on December 31, 1801, almost a year to the day after it had first been sighted among the stars of Taurus. The next night it was independently captured by Olbers. The distance from the Sun, 2.767 astronomical units (an astronomical unit [AU] is the mean distance from the Earth to the Sun), proved to be just right to fit the missing position in the Titius-Bode law. Hegel, needless to say, looked very foolish indeed, though Gauss later remarked that his *Dissertatio* was pure wisdom compared with what he wrote later.

Once again the Solar System seemed complete. Piazzi, to whom the honor was due, named his new planet Ceres Ferdinandea, after the patron goddess of Sicily and its Bourbon prince, Ferdinand IV, though the name was soon shortened to Ceres. The early publications referred to it as a new "major planet," but it soon became clear that it was a very minor member of the Sun's family and never becomes bright enough to be seen without a telescope. In 1802, when William Herschel attempted to measure its diameter with one of his large reflectors, he found it to be a mere 260 kilometers (an underestimate; the modern value is about 1,000 kilometers), so by planetary standards it was little more than a jot.

While he was following Ceres during the months following its rediscovery, Olbers, on March 28, 1802, found another small planet: Pallas. Its distance from the Sun proved to be 2.771 AU. In a letter to Bode, Olbers at once formulated the obvious question: "Did Ceres and Pallas always travel in their current orbits in peaceful proximity or are both part of the debris of a former and larger planet which exploded in a major catastrophe?" As if in confirmation of the latter possibility, Harding, on September 1, 1804, located Juno, while Olbers himself added Vesta on March 29, 1807. Though last discovered, Vesta is the brightest of the four and can even be detected with the naked eye under favorable conditions.

Ceres, Pallas, Juno, and Vesta clearly made up a new class of objects. William Herschel, thinking of their telescopic appearance, coined the name asteroids, which means starlike objects, and this soon came into general use, though in some ways it is unsatisfactory, as obviously there is nothing genuinely starlike about them. After discovering Vesta, Olbers carried on the quest for other asteroids for nine more years. None turned up, and he concluded that whatever fragments might remain were apparently too small to be seen from the Earth.

Nothing more was done until 1830, when a Driesen postmaster, Karl Ludwig Hencke, acquired a small telescope and began his own search. After fifteen years of searching in vain, he was rewarded on December 8, 1845, with the discovery of Astraea. Two years later Hebe fell into his grasp, and he was awarded a pension by the King of Prussia. But this was only the beginning.

The same year, 1847, John Hind of London discovered Iris and Flora. The years 1848 and 1849 saw one new asteroid each, 1850 three, 1851 two, and 1852 no less than eight, which brought the total to twenty-three. The Council of the Royal Astronomical Society confidently predicted that "such a rate of increase among the known members of the Solar System can hardly be expected to continue very long." They could not have been more mistaken; the rate of discoveries increased. By 1857 fifty were known, by 1868 an even hundred.

Seven observers—Hind, de Gasparis, Goldschmidt, Luther, Chacornac, Pogson, and Peters—accounted for sixty-eight of the first hundred discoveries. But the greatest visual observer of all was Johann Palisa, who began his career at Pola, then part of Austria, and later worked at the Vienna Observatory. He found his first asteroid in 1874, and by the time of his death in 1925 he had amassed 125—yet how fleeting is fame. His name is hardly remembered today.

Visual observers like Palisa had no choice but to conduct exhaustive comparisons of areas of the sky against star charts. Any "star" not recorded on the charts immediately became suspicious, the final proof coming from the detection of its motion. Even though the method was highly inefficient, the number of asteroids mounted steadily, and cynics began to refer to them as the "vermin of the skies." One German astronomer quipped that "the value of a new discovery is hardly in proper proportion to the additional work it causes." That was in 1890. The next year Max Wolf of the Heidelberg Observatory introduced a far more powerful method for discovering asteroids: photography.

A telescope can be kept pointed at the same star field, which otherwise would drift out of view owing to the Earth's rotation, by means of a clock drive. If a photographic plate is exposed for several hours, the star images on the plate will register round; an asteroid, on the other hand, since it is moving, will leave a short trail. Wolf's first discovery was the 323d named asteroid, Brucia, on December 22, 1891. By the time of his death in 1932 he had been credited with 232. His assistant, Karl Reinmuth, was even more prolific, with 284.

By June 1991, some 4,848 asteroids had had their orbits worked out, and but for the advent of high-speed computers, keeping track of all of them would long since have become an impossible task. Most, like the original four, hold to the region between Mars and Jupiter, but there are a number of notable exceptions. Number 433 Eros, discovered photographically in 1898 by Carl Witt at Berlin, lies outside the orbit of Mars at aphelion, but at perihelion it may approach within 22 million kilometers of the Earth, as last happened in January 1975. It is an oblong body of which the short axis has been reliably set at 22 kilometers from observations of the asteroid's occultation of the star Kappa Geminorum on the night of January 23, 1975.

A number of asteroids approach even nearer the Earth than Eros. In 1932 Eugene Delporte of Uccle, Belgium, discovered 1221 Amor, whose perihelion lies just outside the Earth's orbit. Only a month later, Reinmuth found 1862 Apollo, which passed only 11 million kilometers from the Earth. Indeed, it skims inside the Earth's orbit, making it the charter member of the Apollo group of Earthgrazing asteroids. Still nearer brushes were made by 2101 Adonis in 1936 (2.4 million kilometers) and by Hermes in 1937 (750,000 kilometers).

In recent years a special search for Apollos (those asteroids that cross the Earth's orbit) has been mounted by Eugene and Carolyn Shoemaker and their colleagues with the 18-inch Schmidt telescope at Mt. Palomar. They have found that Apollos are far more common than used to be thought. Eighty-five were known by the end of 1990, all but fourteen of which had been found since 1970.

One of the most interesting is 1989-FC (a provisional designation, until its orbit can be accurately worked out). It was discovered by Henry E. Holt and Norman Thomas at Palomar. On March 22, 1989, it broke Hermes's half-century-old record by passing within 690,000 kilometers of the Earth, or less than twice the distance to the Moon. An even nearer miss occurred in January 1991 when 1991-BA, discovered by David Rabinowitz at the University of Arizona's Steward Observatory, streaked within only 170,000 kilometers of the Earth. It is the smallest asteroid so far known, measuring at most 9 meters across, which makes it no larger than a small house.

Another interesting Apollo is 1989-PB, discovered by Eleanor Helin, which was studied with the radio telescope at Arecibo, Puerto Rico, as it passed within 2.4 million kilometers of the Earth in August 1989. It proved to be shaped rather like a dumbbell. Such oddly shaped asteroids may form when smaller irregular bodies stick together following gentle collisions. If so, then the existence of simultaneously formed impact craters—the Kara crater in the Arctic and the Usk-Kara crater in the Soviet Union, for example—may be explained by the breakup of such a weakly fused body upon entry into the Earth's atmosphere. The Moon also has such crater pairs—the Messier twins in Mare Serenitatis and Ritter and Sabine in Mare Tranquillitatis are among the best known—and here tidal strains rather than air friction must have produced the breakup.

Since all the known Apollo asteroids have been discovered only when they were very near the Earth, there may be as many as a thousand of them. But where did they come from in the first place? With few exceptions, asteroids stay within the main belt between Mars and Jupiter, but at 2.5 AU there is an unstable zone. Asteroids located there are in a resonance position with Jupiter; that is, they complete exactly three revolutions for each revolution completed by the giant planet. They are, then, regularly disturbed, and as a result their orbits become "chaotic"; that is, the orbital eccentricities vary so much that these asteroids veer across the orbits of other planets, and on making a near approach to one of them may be permanently removed from the system. Thus the region at 2.5 AU has been effectively swept clear of asteroids to create an empty zone, or Kirkwood gap, named after the Indiana astronomer who first investigated the problem in the 1860s. The Apollos are among the former inhabitants that have been removed in this way.

One of the most venturesome of all the Apollos is 1566 Icarus, discovered by Walter Baade in 1949. It not only crosses the orbit of the Earth but those of Venus and Mercury as well. At perihelion, it lies within only 28 million kilometers of the Sun, and its surface then becomes hot enough to melt lead. At aphelion, however, it lies out beyond the orbit of Mars. Besides being unusually elongated, its orbit is sharply inclined to the ecliptic, at an angle of 23°.

A similar asteroid is 3200 Phaethon, which has been known since 1983. It approaches even nearer to the Sun than Icarus, within only 21 million kilometers, and its orbit is almost as sharply inclined. But the most interesting thing about it is that it lies within the Geminid meteoroid stream, which produces the annual meteor shower that occurs each January. Most meteoroid streams follow the orbits of comets and are thought to represent debris they have shed during their approach to the Sun. Before Phaethon was found, the Geminids had been orphans; no parent comet had been identified. It is now clear that Phaethon must be the source of this material, and though at one point it was even suggested that it might be nothing but a burnt-out comet, it looks for all the world like an asteroid, which only goes to show that the distinction between asteroids and comets is to some extent artificial.

As yet no asteroid is known whose orbit lies entirely within that of the Earth, though several, such as 2062 Aten and 2100 Ra-Shalom, both of which were discovered by Eleanor Helin using the 18-inch Schmidt telescope at Palomar, come very close to doing so.

The orbit of an asteroid like Phaethon is such that it could approach the Earth well within the orbit of the Moon. This leads to the far from academic question, What is the chance of a collision with such a body?

The chance, though small, is not zero. William K. Hartmann has estimated that a body 1 kilometer wide strikes an area of the Earth's surface equal to the United States (10 million square kilometers) once every 10 to 20 million years. Such a body would produce a crater more than 100 kilometers across—comparable in size to the lunar crater Copernicus.

It is now well-established that some 65 million years ago, an uncommonly large body—perhaps 10 kilometers across and either an asteroid or a comet—did strike the Earth. The evidence for this comes from the high concentrations of the element iridium found in clay samples from the boundary between Cretaceous and Tertiary rock sediments around the world (the K-T boundary, so abbreviated because the letter C is used for the earlier Cambrian period). Most of the iridium present in the material that originally formed the Earth sank, along with iron and other heavy metals, to the Earth's core, so only traces ought to remain in rocks close to the surface. The concentrations at the K-T boundary are some 20 to 160 times what they ought to be. Clearly the excess iridium had to come from somewhere. As was first realized by Luis and Walter Alvarez and their colleagues at the University of California, Berkeley, in 1979, there was in fact only one possible source: a body from outer space.

In a large impact, a good part of a planet's atmosphere may be blown off, and this may be how Mars lost most of its early mantle. Huge amounts of material, iridium-enriched from the vaporized comet or meteorite, would be borne aloft as a suspension of fine dust, forming a blanket that would screen out much of the solar radiation. In the case of the K-T event, this dusty pall led to subfreezing land temperatures over much of the globe and is believed to have brought about massive extinctions of plant and animal life, including the dinosaurs. Rulers of the Earth for a hundred million years, they suddenly vanished forever from its face.

Such an impact would be bound to leave a large crater. For a while the best candidate seemed to be a 35-kilometer-wide feature near Manson, Iowa, because, though smaller than expected, it was at least of the right age—65.7 million years. However, the main crater has now been identified with reasonable certainty. It is buried beneath the north coast of the Yucatán Peninsula in Mexico and is known as the Chicxulub crater, since the 180-kilometer-wide ring's center lies directly below the town of Puerto Chicxulub. Here, in all likelihood, occurred the death blow that brought about the extinction of the dinosaurs and changed the course of life on Earth forever. Only then did mammals—and ultimately human beings— have their opportunity to take center stage. Though such a dramatic impact would be expected to occur only once in tens of millions of years, much smaller events would still be catastrophic to human life. Even tiny 1989-FC, which is several hundred meters wide, would blast a crater 5 to 10 kilometers in diameter, and at the moment no one can predict just when such an event might take place.

At the opposite extreme from the Earth-glancing asteroids are those which follow unusually distant orbits. In 1906 Max Wolf discovered 588

Achilles, which shares an orbit with Jupiter but which always remains about 60° ahead of it. Later that same year another asteroid, 617 Patroclus, was found in a position 60° behind Jupiter. It turns out that small bodies that always make an equilateral triangle with Jupiter and the Sun will remain in that position more or less indefinitely. The points of the equilateral triangle are called the Lagrangian points, after the French mathematician Joseph Louis Lagrange, who first recognized their theoretical possibility as far back as 1772 but who had no idea that there were bodies that actually met his conditions.

In addition to Achilles and Patroclus, scores of other Trojans (so-called because they were all given names of heroes of the Trojan War) have now been identified at the Lagrangian points of Jupiter's orbit, and of course the same principles of dynamics apply elsewhere in the Solar System. Thus several satellites share orbits around their primary planets, and recently the first Martian "Trojan" has been discovered—an asteroid locked into one of the Lagrangian points of the orbit of Mars.

The Jovian Trojans remained the most remote asteroids known until 1920, when Walter Baade discovered 944 Hidalgo. Its aphelion is located near the orbit of Saturn, and at perihelion it approaches the orbit of Mars. Then in 1977 Charles Kowal found 2060 Chiron with the 48-inch Schmidt telescope on Mt. Palomar. At perihelion, which it next passes in 1996, it crosses just inside the orbit of Saturn. At aphelion it moves out to the orbit of Uranus. At first Chiron was tentatively identified as an asteroid, and it was even suggested that it might be the largest member of an as yet unexplored trans-Saturnian belt. But it is subject to sudden variations in brightness—in 1988, for instance, it was 0.6 magnitudes brighter than it had been in 1986—and recent observations have shown a fuzzy coma, probably made up of dust and carbon dioxide, so it may be a distant comet. If so, however, it is by far the largest known comet, measuring about 180 kilometers in diameter. By comparison, the nucleus of Halley's comet is only 16 by 8 kilometers. In February 1991 a 5-kilometer-wide body, 1991-DA, was discovered by Duncan Steele at Siding Spring Mountain, Australia. Its orbit carries it even farther from the Sun than Chiron and well past the orbit of Uranus, though at perihelion it lies inside the orbit of Mars. Whether it will develop a coma as it approaches the Sun remains to be seen. Currently Eleanor Helin, a leading discoverer of asteroids veering through the Solar System, has launched a systematic survey of the outer Solar System with the 18-inch Schmidt telescope on Palomar Mountain which it is hoped will bring to light other objects of this type.

There remains the question of the physical characteristics of asteroids, a field that for a long time was hopelessly closed. After all, even Ceres has a disk that subtends only about 1 second of arc, which makes the diameters of even the largest extremely difficult to measure directly. E. E. Barnard, using the Yerkes 40-inch refractor, made a series of measures in the 1890s, and for the want of anything better, his estimates remained standard until 1970.

That year Joseph F. Veverka of Cornell University introduced a new method of measuring the diameter of an asteroid that depends on the way its surface polarizes the light it reflects from the Sun. Another method, developed about the same time by David Allen, then a graduate student at the University of Minnesota, involves measuring the brightness of asteroids in the infrared. As Allen knew, the radiation falling onto the surface of an asteroid must be either reflected or absorbed. The portion absorbed goes into heating the surface, but since the surface does not heat up indefinitely, the absorbed radiation must eventually be re-radiated, and so it is, in the infrared. By comparing the brightness of the asteroid in visual and infrared wavelengths, the relative proportions of light that are reflected and absorbed can be worked out, and then all that is needed to determine its diameter is the asteroid's distance from the Earth.

Parenthetically, what prompted Allen's development of his radiometric technique was a view of the asteroid Vesta a few years before with the 12-inch Northumberland refractor at Cambridge University. Barnard had given 380 kilometers for Vesta's diameter, but Allen recalled:

> When Vesta swung into view and settled, I was struck by the fact that it appeared resolved—not quite a disk, but certainly less sharp than the nearby stars. Using Barnard's diameter, I derived 0.3 arc second for its angular diameter—too small to appear resolved in a 12-inch. This experience set me wondering whether Barnard had erred and whether a better diameter-measuring technique could be found.

As it turned out, Barnard's measurement was indeed inaccurate; the currently accepted diameter for Vesta is about 580 by 470 kilometers.

From the recent measurements of asteroid diameters, it has become clear that asteroids fall into three main groups. About three-fourths of them have very dark surfaces that are rich in carbonaceous minerals. These are known as C-type asteroids. On average they reflect only about 5 percent of the light falling on them, and some of them are darker than a blackboard. The next group consists of the stony, or S-type, asteroids. They have surfaces that are very efficient at reflecting sunlight; 44 Nysa, for example, has an albedo of 0.38, which means that it reflects 38 percent of the sunlight falling on it back into space. Finally there are the metallic, or M-type, asteroids, which are made up of nickel-iron.

Ceres is a C-type asteroid, and its diameter is about 1,000 kilometers, which makes it considerably larger than Barnard's estimate of 720 kilometers. Vesta and Pallas are other examples of the type. The diameter of Pallas is especially well known, having been measured at 517 kilometers during its occultation of the star 1 Vulpeculae on May 29, 1983. However, at least ten asteroids are larger than Juno, which at only 240 kilometers is among the four brightest asteroids only because it is of the S-type and thus has a highly reflective surface. Those that surpass it are far from household names. They include, for example, 10 Hygeia (430 kilometers), 511 Davida (340 kilometers), and 704 Interamnia (330 kilometers).

Though Olbers once thought them to be remnants of a planet that broke apart, the asteroids instead represent a planet that never formed, having been prevented from doing so by Jupiter, whose perturbing influence produced, in the zone just inside its orbit, a kind of Cassini Division in the Solar System. They are thus, like comets, leftover examples of the planetesimals out of which the planets were fashioned, though the asteroids are in far from pristine condition. They have been fragmented by innumerable collisions and blasted by micrometeorites and the particles of the solar wind.

So far, one main-belt asteroid has been examined at close range: 951 Gaspra, which the Galileo spacecraft examined from a distance of 16,000 kilometers on October 29, 1991. Gaspra is an irregularly shaped body measuring 20 by 12 kilometers. Thus it is about the same size as Mars's moon Phobos. Indeed, in appearance it closely resembles the two tiny Martian moons, which is hardly surprising, given that they are themselves probably captured asteroids. Gaspra's irregular shape suggests that it is a fragment from the collision of two larger bodies. From the relative lack of small impact craters on its surface, it is believed that this collision in all likelihood occurred no more than several hundred million years ago.

For amateurs with specialized equipment, there is useful asteroid work to be done. For example, using a photometer, variations in the asteroids' brightness can be measured, which leads to information about their rotation periods. Most, though, will be content simply to capture them among the stars—and in even a 3-inch telescope, well over a hundred are within reach. It is always satisfying to ferret out one of these little worlds as it wheels its endless path around the sun.

CHAPTER NINE

Jupiter

Jupiter orbits majestically around the Sun at a mean distance of 778 million kilometers, completing each circuit in a period of twelve Earth years. Because the planet moves relatively slowly, the Earth catches up with it and brings it to opposition at intervals of roughly one year. Moreover, its highly reflective cloud deck and enormous size—it is the largest planet, with a volume 1,300 times the Earth's— mean that it shines brightly in the sky even far from opposition. It is thus unlike Mars, which shines splendidly at opposition once every two years but then fades rapidly as it withdraws from the Earth. Even when farthest away, Jupiter has a disk that is larger than that of Mars at its closest. It can therefore be studied to advantage for months before or after the date of opposition.

In Jupiter we see a world of a completely different order from the Earth and the other "terrestrial" planets. It is the first and greatest of the gas giants, which reign over the outer reaches of the Solar System. There is no solid surface to stand on, only clouds that become thicker and hotter as one descends into the planet's depths.

Even a small, 20× telescope suffices to show Jupiter's flattened disk. Measured through the equator, the diameter is 142,800 kilometers; through the poles it is only 134,200 kilometers. The polar flattening is a result of the centrifugal force generated by the rapid Jovian rotation, at just under ten hours the shortest of all the major planets.

On Jupiter we are clearly dealing with meteorology rather than geology, and thus a few basic principles should be established at the outset. Assuming an orthodox axial tilt, gas near a planet's equator is heated more than that near the poles, and in this way cells of circulation between different zones of latitude are set up. On the Earth, as an air mass leaves the equator for the poles it is deflected eastward relative to the ground because of the Earth's rotation. Conversely, an air mass moving toward the equator is deflected in a westward direction. These deflections are due to the so-called Coriolis effect, which is responsible for producing a pattern of alternate westerly and easterly currents. Thus there are the westerly trade winds between latitudes 5° and 30°N and S, the temperate easterlies between 35° and 50°, and the polar westerlies at still higher latitudes.

The predominantly yellowish clouds of Jupiter are banded, with light zones and dark belts. These were first seen by Galileo's pupil Evangelista Toricelli in 1630. The main ones, at least, tend to remain reasonably stable over time, and they bear witness to something similar to the alternate westerly and easterly wind currents of the Earth, though of course Jupiter's wind currents are bound to be much stronger, given its enormous size and rapid rotation. On the other hand, other processes may be generating the planet's zonal winds. According to one theory, as gas rises by convection from the warm interior, it is sheared by the centrifugal force of the planet's rotation into a series of differentially rotating cylinders, one nested within the other. According to this view, wherever the surface of one of these cylinders pokes through the cloud tops, it gives rise to a zonal current.

Already in the seventeenth century, Christiaan Huygens recognized that Jupiter was a "windy planet." G. D. Cassini went a step further and showed that spots near the equator have rotation periods five minutes shorter than those near the poles, which we now know is due to the existence of the powerful Equatorial Current. Moreover, in 1665 Cassini made out a particularly impressive spot, located in what we now refer to as the South Tropical Zone. He called it the Eye of Jupiter. The same spot had been sketched a year earlier by Robert Hooke of the Royal Society, and indeed it was observed from time to time until 1713, when Cassini's nephew Jacques Maraldi recorded it for what proved to be the last time for over a century.

Inexplicably, Jupiter, which is among the most rewarding objects for small telescopes, was rather neglected during the eighteenth century. During the nineteenth century the observational situation gradually improved. Gruithuisen, for instance, recognized the "wisps" that often appear in the Equatorial Zone, and Heinrich Schwabe of Germany and W. R. Dawes, William Lassell, and Laurence Parsons, the fourth earl of Rosse, all of England, made valuable observations. However, it was not until 1878 that Jupiter really became the center of attention.

That year the feature known ever since as the Great Red Spot erupted into prominence on the planet's surface. The spot's color was described as brick red, and the dimensions of its oval outline reached, according to the English observer W. F. Denning, 40,000 kilometers by 13,000 kilometers, which made its area three times that of the Earth. It so dominated the Jovian scene that whenever the side of the planet on which it was located was turned toward the Earth, it could not be missed.

The Red Spot was not new; only its prominence was. Indeed, only a year before, observers in New South Wales had described it as the "pink fish." Rosse had seen it several times during the early 1870s, and Dawes once in 1857, but it had attracted scant attention. Going back still further, Schwabe in 1831 had recorded the notch or "Hollow" it carves out of the adjacent South Equatorial Belt, which remains visible even when the Red Spot itself is invisible. Despite the lack of an intervening record, there is no reason to doubt that it is the same spot as that seen by Hooke and Cassini more than three centuries ago.

After its rise to prominence in 1878–79, the Red Spot remained brick red only until 1882. Then it began to fade. For the next several years it hovered near the brink of invisibility, only to revive again in 1891. Since then it has been dark at certain times and faint at others. This is, indeed, its characteristic behavior. From the mid-1960s through most of the 1970s it was an easy object to observe in small telescopes, often appearing distinctly salmon pink. It became faint during the 1980s, rose briefly to prominence in 1989–90, began to fade again in August 1990, and by June 1991 had once more become inconspicuous.

We now know that the Great Red Spot is a huge vortex in the planet's atmosphere, but in the 1870s its nature was far from certain, and some rather unusual theories were put forward to account for it. Camille Flammarion wondered whether it might be a continent in formation, while W. F. Denning regarded it as a gaping rent in the clouds through which, he wrote, "we saw the dense red vapors of his lower strata, if not his actual surface itself." Denning's remark alludes to the then widely held belief that Jupiter was a kind of half-fledged star, a world arrested in development between a planet and a sun. After all, Jupiter is so remote that the intensity of the solar radiation there is only 1/27th of what it is on the Earth, but its clouds are always in vigorous motion. To support such activity it seemed that Jupiter needed some energy source besides the Sun—in other words, that it needed to generate some of its own heat. The English astronomer Richard A. Proctor summed up the matter in 1870:

That enormous atmospheric envelope is loaded with vaporous masses by some influence exerted from beneath its level. Those disturbances which take place so rapidly and so frequently are the evidences of the action of forces enormously exceeding those which the Sun can by any possibility exert upon so distant a globe. . . . [W]e seem led to the conclusion that Jupiter is still a glowing mass, fluid probably throughout, still bubbling and seething with the intensity of the primeval fires, sending up continuously enormous masses of clouds, to be gathered into bands under the influence of the swift rotation of the giant planet.

Amazingly, this hunch turned out to be quite right. In 1969 infrared observations showed that Jupiter does indeed emit twice as much heat as it receives from the Sun.

Following the Great Red Spot's rise to prominence, Jupiter finally began to receive the attention it deserved, and an especially elegant piece of work was carried out by an English amateur, Arthur Stanley Williams. He was born at Brighton in 1861 and commenced a long series of observations of Jupiter when he was seventeen, about the time the Red Spot was taking center stage. Later he became a solicitor by profession, but Jupiter remained his ruling passion—that and yachting, in which he also made a mark. Few have contributed as much as he to our understanding of the planet, yet he used only a 6 1/2-inch reflector right up to the final year of his life. He died in 1938.

Williams's chief goal was to make a detailed study of the circulation patterns of Jupiter, and his methods were disarmingly simple. He recorded to the nearest minute the time at which each Jovian feature crossed the planet's central meridian, and by following a given feature through several transits he was able to work out its rotation period very accurately. In this way the feature served, in effect, as a probe of the Jovian winds, allowing him to map out the various currents. In 1896 he published his classic paper "On the Drift of Material in Different Latitudes of Jupiter," in which he announced the existence of nine distinct currents. Building on this work, later observers added still other currents until today no less than nineteen have been recognized between latitudes 60°N and s.

The distinctive belts and zones of the planet, which can be made out even with a 2-inch telescope, show a rough correlation with the currents. This is, of course, only to be expected, but the agreement is far from exact, and the real significance of the belts and zones lies in the fact that they distinguish different levels of clouds. The bright zones are cloud features that are relatively higher than the dark belts.

Because on Jupiter we are looking at a "surface" that consists entirely of clouds, there are obvious problems with defining a zero meridian of longitude—the Jovian equivalent of Greenwich. For convenience, the planet is divided up into two zones of longitude. Spots within about 9 degrees either side of the equator have an average rotation period of 9 hours, 50 minutes, 30 seconds. This is the region of the westerly Equatorial Current, first recognized by Cassini so long ago, and is described as System I, which by definition has a period of 9 hours, 50 minutes, 30.003 seconds. Higher latitudes fall within the domain of System II, whose period is 9 hours, 55 minutes, 40.632 seconds, which was the mean rotation period of the Great Red Spot in 1890–91. Dividing up the planet in this way gives a reasonably good idea of the average relative drift of features lying in the two regions, though obviously it is an oversimplification and individual spots have their own independent motions.

As if all of this were not complicated enough, there is yet another system of longitudes, System III, whose period is 9 hours, 55 minutes, 29.7 seconds. This is the time it takes the solid core of the planet to rotate. It was quite unknown until the 1950s, when it was discovered by radio astronomers.

Ever since Williams's time, many of the leading students of the planet have been amateurs, and planetary astronomy owes them an enormous debt. The casual view reveals little more than a banded disk, but as the eye makes a more careful study, ever more minute features dart across the threshold of perception. The edges of the belts are less smooth and regular than appears to the casual glance; instead they are notched, broken, and serrated. Plumes or loops may sweep out from them into an adjoining zone. Then there are the various light and dark spots, some of which are best described as evanescent, while others, like the Great Red Spot, endure for many centuries. Complex as the detail visible in even modest instruments may be, each increase of resolution reveals intricacies within intricacies. Jupiter is, in a word, a Chinese puzzle of a planet.

At this point it may be useful to review the main Jovian features in a more systematic way. Needless to say, this account is only meant as a general guide, but it will give at least some idea of the complexity of this fascinating planet.

Starting at the north pole (see fig. 9.1), there is the large dusky hood known as the North Polar Region (NPR). Several variable belts and zones

FIGURE 9.1. Jupiter, photographed by Voyager 1 in February 1979.
The nomenclature for the various belts and zones is indicated.
(NASA photograph)

lie between the NPR and the first main Jovian belt, the North Temperate
Belt (NTB). This belt is remarkable in that spots at its southern edge have
had the shortest rotation periods ever observed on the planet—9 hours,
49 minutes. They are swept along by a strong westerly jet located there.

South of the NTB are the North Tropical Zone (NTrZ) and the North
Equatorial Belt (NEB). During most of the twentieth century the NEB has
been the most prominent Jovian belt, and its appearance is very characteristic. Its northern edge—which together with the NTrZ lies in the
easterly North Tropical Current (System II)—is often marked with elongated brownish cloud formations descriptively referred to as "barges."
The southern edge of the NEB lies in the Equatorial Current (System I) and
forms indentations and knoblike projections from which plumes or wisps
often sweep into the Equatorial Zone.

The Equatorial Zone (EZ) is the broad girdle of the planet, occupying fully one-eighth of its surface area. This is the domain of the strong westerly Equatorial Current, where the wind speeds are 540 km/hr. The EZ is one of the most active areas on the planet, and here some of my own observational notes may be of interest. In 1966–67 I found the EZ to be brilliant white. Thereafter a definite darkening set in, and by 1972 the color could best be described as tawny. In 1973 several whitish plumes appeared, emerging from small whitish spots on the NEB. Two were especially prominent, having a somewhat faster rotation period than other features in the zone. Two years later the whole region was filled with regularly spaced brownish plumes, giving it a beautifully festooned appearance. When Voyager 1 arrived at the planet in 1979, there were no less than thirteen plumes. By 1987–88 all of them had disappeared. The EZ was once more brilliant white, and it remained so until October 1989, when a sudden darkening took place in its northern half.

Bounding the EZ on its southern flank is the South Equatorial Belt (SEB), often double and usually somewhat inferior to its counterpart, the NEB. However, between 1982 and 1989 it was the most prominent belt on the planet. In May 1989 it began to fade and actually remained invisible for more than a year, the entire region being filled with whitish clouds. In August 1990 a thin brownish band began to encircle the planet, and within a month the SEB had returned.

The whole region of the SEB is the most variable on the disk, and periodic fadings and revivals of the belt are a prominent feature of the planet's meteorology. The pattern has usually been as follows. After a period of indistinctness, dark spots appear in the latitude of the SEB, moving retrograde with respect to the planet's rotation—these spots mark the beginning of an "SEB disturbance." Soon other spots appear, this time moving in the opposite direction, and after several weeks the two groups of spots meet up with one another. Before long the whole SEB becomes filled with numerous spots, wisps, and other complex details, giving it a highly disturbed appearance and signaling the return to prominence of the belt. The interval between disturbances is three to four years, with the events of 1919, 1928, 1943, and 1975 being particularly violent.

The explanation for the cycle of events is as follows. When the SEB appears whitish, it is because a deck of high ammonia cirrus clouds is present. A "disturbance" consists of a bubble of hot gas originating at one of three "hot spots" deep within the planet. As it rises, a bubble gives up its heat through condensation and produces a spectacular eruption of brownish clouds that eventually envelops the whole region (thus bringing about the revival of the belt).

The South Tropical Zone (sTrz) is ruled by the Great Red Spot (GRS), but apart from the Red Spot itself, the most notable feature was the famous South Tropical Disturbance (sTrD), which during most of its lifetime (from 1902 to 1940) appeared as a broad shaded region in the sTrz whose interactions with the Red Spot, which had a slightly different rotation period, were noteworthy. Comparatively minor disturbances occurred from 1955 to 1958 and from 1978 to 1983.

The South Temperate Belt (sTB) has been home to three white oval spots, designated BC, DE, and FA, which first appeared in 1939. Except for their color, they are of the same character as the Great Red Spot, and after it were the longest lived features on the planet. During the 1970s and 1980s I found them very prominent, but in 1988–89, when the sEB itself was faint, they merged with the general background of whitish clouds and often could not be seen at all.

In the extreme south of the disk there are several variable belts and zones, which finally become indistinguishable from the dusky South Polar Region (sPR), the southern counterpart of the NPR.

Even this brief description should suggest that the planet is one of great variety and sometimes of violent change. Moreover, much of what I have described here lies within reach of modest instruments—to a large extent I have relied on my own observing notes in drawing up this account, and I make no claim to having exceptional skills as an observer.

A new era in the exploration of Jupiter began when the Pioneer 10 spacecraft passed by the planet on December 4, 1973. It was followed a year later by Pioneer 11. Their images showed the planet in a pre-sEB disturbance state, with resolution considerably better than had ever been achieved from the Earth. However, their results were overshadowed by those of the far more ambitious Voyager spacecraft.

The Voyager missions were planned to take advantage of the unusually favorable alignment of the outer planets that occurred in the late 1970s, and Jupiter was the first destination of the two spacecraft. Voyager 1 swept within 278,000 kilometers of the Jovian cloud tops on March 5, 1979, and Voyager 2 within 650,000 kilometers on July 9 of the same year.

The planet was much more active in 1979 than it had been during the Pioneer flybys of 1973–74. Everywhere there were wild, convoluted cloud patterns. The Equatorial Zone was filled with regularly spaced plumes, and the Great Red Spot, rolling between two opposite-flowing streams, stirred the clouds of the South Equatorial Belt into a riot of colorful swirls.

In the shear zones between easterly and westerly currents, the Voyager cameras recorded an abundance of small eddy activity. The eddies begin when cells of warm air rise from the depths by buoyancy, acquiring vorticity through the Coriolis effect. In general, the eddies are highly unstable under the turbulent Jovian conditions, lasting only a day or two before being torn apart by the violent zonal jets, but as they break up they surrender their energy to the jets. Thus, by cannibalizing the small eddies, the jets are able to maintain their own stability, and indeed they have changed little in position or wind speed since the days of Stanley Williams.

The largest eddy and most successful cannibal of all is the Great Red Spot, which circulates between two opposing jets with a period of six days. It is an elevated high-pressure area, and its cloud tops are much cooler than the other Jovian clouds, as they loom several kilometers above them. Material from below is being channeled upward, and just as happens with the small eddies, the ascending gas acquires vorticity. In the southern hemisphere, this results in counterclockwise or anticyclonic motion. Not only the Great Red Spot but also the white ovals of the South Temperate Belt are large anticyclonic eddies, and the Voyagers revealed many smaller features of the same general type.

How do features like the Red Spot and the white ovals avoid breaking up on such a turbulent planet as Jupiter? They do roll smoothly between the jet currents on either side, but even so they would run down eventually if they did not continually feed on smaller eddies. The upshot of this is that, although on a small scale the planet's cloud patterns appear highly chaotic, out of them a large-scale order emerges.

In composition, the bright clouds of the zones are ammonia cirrus. They occupy the region of the atmosphere where the temperature is lowest, $-113°C$. At higher levels there is a warming rather like that which occurs in the Earth's stratosphere. Though on the Earth the ozone layer is responsible for stratospheric warming, on Jupiter a hydrocarbon smog plays this role.

Parenthetically, it should be mentioned that, under equilibrium conditions, ammonia cirrus clouds would appear perfectly white. The intense colors observed on Jupiter must therefore be due to disturbances of chemical equilibrium by charged particles, lightning, or rapid vertical mixing through layers of different temperature. Various polymers of elemental sulfur are thought to be important in producing the rich yellows and browns of the clouds, while elemental phosphorus, P_4—presumably formed from phosphine, PH_3, through a series of chemical reactions—has been invoked as a likely ingredient responsible for the sometimes intense hues of the Great Red Spot.

The atmospheric pressure at the ammonia cirrus layer is about equal to that at the Earth's surface; at greater depths, the pressures and temperatures steadily rise. The clouds of the dark belts are made up of ammonium hydrosulfide and are formed in a region about 100 kilometers below that of the ammonia cirrus layer. The temperature there is about 100°C warmer than at the upper cloud tops.

It is believed that below the ammonium hydrochloride cloud layer lies a layer of cirrus clouds of water-ice crystals and below this a layer of liquid water itself. At still greater depths, the planet's structure is determined chiefly by the behavior of hydrogen under different conditions of temperature and pressure. The planet is 90 percent hydrogen by volume, which is only to be expected, because hydrogen is by far the most common element in the universe, with helium a distant second. The Voyagers found that the ratio of hydrogen to helium on Jupiter is the same as that on the Sun— nine hydrogen atoms to each helium atom, which reflects the relative cosmic abundance of these elements.

Only in the outermost shell of the planet, to a depth of perhaps a thousand kilometers, does hydrogen behave like a gas. At greater depths it liquefies, while at a depth of about a third of the way down to the planet's center there is another phase transition where hydrogen begins to behave like a metal, that is, a conductor of electrons. The rotation of this metallic hydrogen core produces electric currents, and this in turn leads to the generation of a powerful magnetic field. At the very center of the planet there is believed to be a small rocky core. The temperature there is about 30,000°C, and the pressure is about 100 million times that of the Earth's atmosphere at sea level.

Charged particles, or ions, emanating from the Sun make up the "solar wind." As they stream by Jupiter they become trapped in its magnetic field, and since the magnetic field rotates with the metallic hydrogen core of the planet, the charged particles rotate along with it. In this way they may be accelerated to velocities approaching the speed of light, and as they are accelerated they emit electromagnetic radiation, of which the most intense bursts are found in the radio range. In fact, it was by recording such bursts from Jupiter that radio astronomers in the 1950s first established the rotation of the core of the planet.

The region around a planet where charged particles are trapped in the magnetic field is known as the magnetosphere, and Jupiter's is so vast that, were our eyes sensitive to radio emissions, it would appear to subtend twice the apparent diameter of the full Moon. All of the planet's satellites travel within it, and Io in particular is involved in remarkable interactions with it whereby the radio bursts from Jupiter are intensified whenever Io comes to certain positions in its orbit. Electrical discharges between the planet and Io may explain this effect.

The four largest satellites of Jupiter were discovered by Galileo in the winter of 1610. As mentioned earlier, he named them the Medicean Stars after his patron, Cosimo de' Medici II, but their modern names—Io, Europa, Ganymede, and Callisto, after the mythological lovers of Jupiter—were proposed by Simon Marius, a German observer who disputed with Galileo over priority in their discovery. He did not publish his results until 1614, however, and by then it was, as the Jesuit Christoph Scheiner noted, "in vain and too late." Even a 6-inch reflector begins to show the four "Galileans" as tiny disks, and it is always fascinating to watch the transits of the satellites and their shadows across the planet (fig. 9.2), the eclipses that occur when the satellites pass into Jupiter's shadow, and the occultations when they pass behind its disk.

In small telescopes the shadows appear inky black throughout. They were apparently seen first by Riccioli in 1643, but Cassini made more detailed observations. On the other hand, each satellite has its own distinctive appearance in transit. Io and Europa are relatively bright and tend to be lost when seen against bright clouds, but they stand out well in projection against duskier cloud features. Ganymede and Callisto, on the other hand, are so dark that at times they appear indistinguishable from shadows.

The dimensions of the Galileans make them planet-sized worlds. Ganymede, at 5,280 kilometers, is not only the largest of the Galileans but the largest satellite in the Solar System. Callisto is 4,820 kilometers in diameter; Io, 3,630 kilometers; and Europa, 3,125 kilometers. Callisto and Ganymede are half rock and half water ice. Europa and Io are almost entirely rock.

The first detailed views of the Galilean moons were obtained by the Voyager spacecraft. Callisto, the darkest of the Galileans, has a surface of dirty ice and is the most heavily cratered body in the Solar System (see fig. 9.3). Its crust is made up of water ice, which has a lower tensile strength than rock. Thus, despite its battered condition, its surface is far less dramatically sculpted than Mercury's or the Moon's, and the terminator appears remarkably smooth. Although many of the larger impact features

FIGURE 9.2. This is one of the most striking images captured by the Voyager 1 mission to Jupiter. It was taken on February 13, 1979, when the spacecraft was still 20 million kilometers from the planet. Two Galilean satellites are shown in transit: Io, projected against the Great Red Spot, and Callisto. (NASA photograph)

have been all but obliterated by ice flows, Callisto has several conspicuous multiringed basins. The dominant one, near the equator, is known as Valhalla and is surrounded by broken ridges extending to a distance of 1,500 kilometers from its center. Again, however, there is a curious lack of vertical relief, and the multiringed basins of Callisto are well described as ghost basins.

Like Callisto, Ganymede also has an icy surface, but its structure is much more complex (see fig. 9.4). There are primitive regions, dark and heavily cratered, which have been given such names as Galileo Regio, Marius Regio, Perrine Regio, Barnard Regio, and Nicholson Regio. This older crust has broken up into plates separated by regions of bright grooved terrain known as sulci.

Europa (fig. 9.5) is a world unto itself. It has the smoothest surface of any of the Galileans, indeed of all the bodies in the Solar System. It is generally bright, but there are strange mottlings and a system of complex lineations that, except for the absence of relief, look remarkably like the cracks seen in the large ice fields of the Earth's polar regions. It may well be

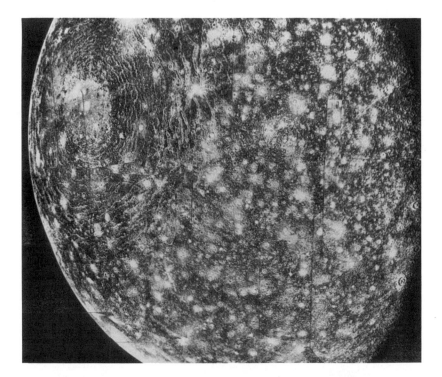

FIGURE 9.3. A photomosaic of Callisto made from Voyager 1 images obtained on March 6, 1979, from a distance of about 200,000 kilometers. The satellite is heavily cratered, and many of the craters have bright ray systems. Nevertheless, the overall appearance is very different from that of the Moon, owing to the relative lack of relief. The outstanding feature is the large basin Valhalla, which dominates the left side of the photomosaic. Its numerous concentric rings were formed by shock waves produced by the impact. (NASA photograph)

that the distribution of these lineations is controlled by a system of deeper fractures within the satellite's water-ice crust. This crust probably extends to a depth of about 50 kilometers and is believed to lie atop a liquid water layer, kept from freezing by tidal heating—but of this more presently.

The existence of an underlying ocean on Europa may also help to explain the so-called arcuate ridges, low wrinkles in the crust that repeat at intervals of a hundred to several hundred kilometers. These ridges seem to have formed when the surface area of the satellite expanded by several percent, and this most likely occurred when the underlying ocean partially froze (remember, water ice occupies a larger volume than liquid water).

We turn finally to Io, surely one of the strangest worlds in the Solar

System (see fig. 9.6). Voyager 1 found no less than eight active volcanoes during its March 1979 flyby. The largest volcano, and also the first discovered, is Pele, named after the fiery Hawaiian volcano goddess. The others have been named Prometheus, Loki, Volund, Amirani, Maui, Marduk, and Masubi. When Voyager 2 arrived four months later, Pele had become quiet, but the others were still erupting. Indeed, Prometheus and Loki had become even more active. Ever since, Loki's activity has continued to be monitored from Earth, and several new volcanoes have been discovered. The gas and dust from Io's volcanoes are ejected at speeds of up to 0.5 to 1 km/sec. This is much higher than in even the most powerful terrestrial volcanoes, such as Pinatubo and Krakatoa, and it suggests that Io's eruptions more nearly resemble terrestrial geysers. Sulfur gas fuels the explosions from Pele, while sulfur dioxide does the same for Prometheus and Loki. Because of the force with which the gas and dust are released and the satellite's low surface gravity (about one-sixth of the Earth's),

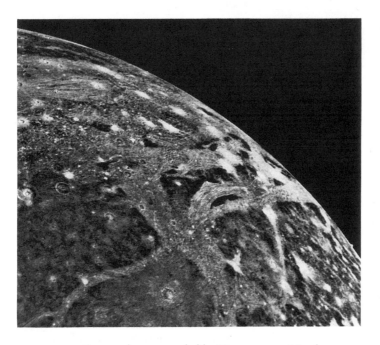

FIGURE 9.4. Ganymede, as recorded by Voyager 1 on March 5, 1979, from a distance of 253,000 kilometers. Its surface shows numerous impact craters, some surrounded by bright ray systems, and bands of light grooved terrain known as sulci. (NASA photograph)

FIGURE 9.5. Europa, photographed by Voyager 2 on July 9, 1979, at a range of 240,000 kilometers. The icy surface is remarkably smooth, with few impact craters, and it displays a complex system of linear features, which, though they look like cracks, do not represent significant differences in relief. The arcuate ridges, described in the text, are not visible in this image. (NASA photograph)

umbrella-like plumes reach heights of hundreds of kilometers and strew material over great distances, so the entire surface must be covered with a new layer of dust perhaps a millimeter thick over the course of a year. Over the course of geologic time, the moon must almost have turned itself inside out.

Though coming as a surprise to most astronomers, the existence of active volcanoes on Io was not entirely unexpected. In a paper published just three days before the Voyager 1 flyby, Stanton Peale, Patrick Cassen, and Ray Reynolds suggested that Io's interior might well be molten owing to tidal interactions with Europa and Ganymede. In their own words: "Calculations suggest that Io might currently be the most intensely heated terrestrial-type body in the solar system. . . . One might speculate that

FIGURE 9.6. A composite of several images of Io made by
Voyager 1 on March 4, 1979. Io's surface has been—and
continues to be—fashioned by explosive eruptions and
lava flows. (NASA photograph)

widespread and recurrent volcanism would occur, leading to extensive
differentiation and outgassing."

All of the Galileans have captured rotations with respect to the giant
planet, and their orbits are almost exactly circular. Whenever Io passes
Europa or Ganymede, it gets a tug from them that pulls it slightly out of
line. Jupiter, which never relinquishes its magisterial grip on its vassal,
then pulls it back again. This planetary tug-of-war generates friction in
Io's interior, melting the rock. The lighter, more volatile elements have
long since been forced to the surface and boiled away. This has left the
outermost layers of the crust rich in sulfur compounds.

At the low temperatures found at Io's distance from the Sun, yellow
sulfur ought to appear nearly white. When heated, however, yellow sulfur
becomes orange-red and more viscous. At still higher temperatures it
becomes a black liquid. All of these colors are found on the surface of Io.

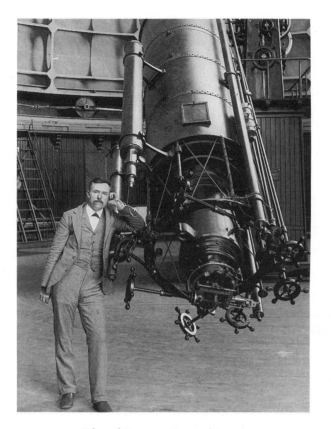

FIGURE 9.7. Edward Emerson Barnard (1857–1923)
(Courtesy Mary Lea Shane archives of the Lick
Observatory)

Loki in particular contains a black lake of liquid sulfur some 250 kilometers across within which float icebergs of solid sulfur.

The material spewed from Io's volcanoes has an interesting subsequent
history. Some of it, of course, falls back to Io, but some of it also escapes
and, colliding with ions trapped in Jupiter's magnetic field, has electrons
knocked off and becomes ionized. This ionized gas forms a large ring
surrounding Io known as the plasma torus.

The twelve remaining Jovian satellites are insignificant little worlds. The
fifth satellite, Amalthea, was discovered in 1892 by E. E. Barnard using
the 36-inch refractor at Lick Observatory. It was the last satellite in the
Solar System to be discovered visually and is now known to be an oblong

object measuring 270 by 155 kilometers. The other satellites were all discovered photographically, the last three by the Voyager spacecraft.

For a long time it was customary simply to refer to the satellites by Roman numerals assigned in order of their discovery, but they have now received official names. Starting with the innermost satellite and working outward, they are Metis (XVI), Adrastea (XIV), Amalthea (V), Thebe (XV), Io (I), Europa (II), Ganymede (III), Callisto (IV), Leda (XIII), Himalia (VI), Lysithea (X), Elara (VII), Ananke (XII), Carme (XI), Pasiphae (VIII), and Sinope (IX). The orbits of the outer four are retrograde, while Metis and Adrastea are embedded in a thin Jovian ring, another surprising Voyager discovery. It consists of a diffuse belt of material about 6,000 kilometers wide. The ring was exceedingly faint even in the Voyager 1 images, though in views from Voyager 2 taken from behind the planet, it brightened up, proving it to be made up of fine dust.

Aside from the Galileans, there is little in this whole retinue to interest the amateur, and only Amalthea and Himalia are more than 100 kilometers across. Himalia is actually easier to see, as Amalthea is always badly swamped by the planet's glare. At magnitude 15, Himalia should be within reach of observers with very large apertures.

Following the Voyager missions of 1979, Jupiter and its satellites once more became the purview of Earth-based telescopes, and amateurs resumed their important role of keeping the record up-to-date. All of this will change again when the Galileo spacecraft goes into orbit around the planet on December 7, 1995, assuming, that is, that its main antenna can be unstuck by then. Needless to say, there is no way that amateurs with their modest telescopes can hope to compete with the high-resolution views obtained from spacecraft. Nevertheless, there is nothing more satisfying than to record for oneself the developments on this turbulent planet and to watch the graceful minuet of the four large moons. Except for Mars on the rare occasions when it is unusually near the Earth, Jupiter is easily the most rewarding planet for those with modest telescopes. It is a world of infinite variety, and every opposition holds the promise of finding something new and exciting.

CHAPTER TEN

Saturn

Saturn, the most remote of the planets known to the ancients, appears to the naked eye as a "star" of yellowish hue. It wanders leisurely among the background stars, completing an orbit in 29 1/2 Earth years. Like Jupiter, it is flattened at the poles, having a diameter of 120,000 kilometers at the equator and 108,000 kilometers at the poles.

To the naked eye, Saturn is bright but far from spectacular, and there was nothing to prepare Galileo for the remarkable appearance he witnessed when he turned his small telescope toward it in July 1610. As he announced to the Grand Duke of Tuscany, "I have discovered a most extraordinary marvel. . . . The fact is that the planet Saturn is not one alone, but is composed of three, which almost touch one another and never move nor change with respect to one another." Galileo assumed, naturally enough, that the two lesser bodies were satellites, but if so they were strange satellites indeed, as two years later they had completely disappeared.

In time the "satellites" returned, and astronomers desperately tried to explain the sequence of events. Gilles de Roberval of France suggested that the equator of Saturn was a torrid zone giving off vapors, while a Jesuit, Honoré Fabri, promoted a theory that explained the changing appearance as due to the periodic passage of two dark satellites in front of two bright ones.

The true state of affairs was finally discerned by Christiaan Huygens. Using an improved refractor of his own design that magnified about 50×, he began to suspect the existence of a ring around the planet as early as

1655. To establish his priority he published an anagram that when rearranged announced that the planet is surrounded "by a thin, flat ring, nowhere touching, and inclined to the ecliptic." Huygens's full theory, together with the solution of the anagram, was published in his book *Systema Saturnium* in 1659. There Huygens showed that every fourteen or fifteen years the Earth passes through the plane of the ring. This had been the case in 1612, but because of the ring's extreme thinness, it had disappeared in the small telescopes of the day (see fig. 10.1). At other times the ring can be tilted toward the Earth by an angle of as much as 28°, which is the inclination of Saturn's orbit to the Earth's. Depending on just how the rings are presented, the brightness of the planet may vary by a factor of two and a half.

What the first observers saw as one ring is actually divided into two rings by a jet-black line. This was discovered by G. D. Cassini at the Paris Observatory in 1675. The outer ring is designated Ring A, and the inner, by far the brighter of the two, is Ring B. The Cassini Division is only 0.5 seconds of arc in breadth, yet the keen-eyed Italian was able to make it out with only a 2 1/2-inch refractor.

Except for Cassini, who suggested that the rings were probably made up of "a swarm of tiny satellites," the early astronomers assumed that they were solid. But how could such large rings, if solid, remain intact? The question was investigated mathematically by Pierre Simon de Laplace of France, who in 1785 came to the conclusion that a solid ring as large as Saturn's was an impossibility; it would be bound to break up, and he realized that even two solid rings would be unstable. Instead, Laplace argued for many solid ringlets, each extremely narrow, and to bolster his claim he recalled that the English instrument maker James Short, who had died in 1768, had once actually reported seeing many divisions in the rings.

The great observer William Herschel was frankly skeptical of Laplace's theory. He reluctantly accepted, after many observations, that the Cassini Division was a true gap—this meant that there had to be at least two solid rings—but he could never bring himself to believe in what he called Laplace's "narrow slips of rings." Perhaps, Herschel suggested, the ring system could break up into many narrow ringlets and then reunite into two, but he had little faith even in this, as shown by his last words on the subject:

> The mind seems to revolt, even at first sight, against an idea of the
> chaotic state in which so large a mass as the ring of Saturn must

FIGURE 10.1. Saturn, showing phenomena of the edgewise
rings. In the drawing at top, the rings are completely
invisible, though the black shadow they cast is prominent,
cutting across the middle of the planet. At bottom, the rings
have reappeared. Note their threadlike appearance to the
right of the globe. (Drawings by Richard Baum, based on
observations with a 4 ½-inch refractor at 186×. Top: March
3, 1980, 25:10 UT; bottom: June 5, 1980, 21:00–23:00
UT)

needs be, if phenomena like these can be admitted. Nor ought we to
indulge a suspicion of this being a reality, unless repeated and well-
confirmed observations had proved, beyond a doubt, that this ring
was actually in so fluctuating a condition.

Herschel himself had only once seen anything resembling a new division
in Saturn's rings, a black "list," or linear marking, that appeared on one
side of Ring B near its inner edge in June 1780. But later observers
reported greater success. Thus in 1837 J. F. Encke at Berlin reported a

shading in Ring A that divided that ring into two nearly equal parts, and the following year Father Francesco De Vico noted faint divisions in Ring B. It seemed that Laplace's theory had been confirmed.

Doubts began to arise in 1850, however, when a dusky inner ring was discovered by W. C. and G. P. Bond at Harvard and independently by W. R. Dawes in England. Dawes's friend William Lassell referred to it picturesquely as the crêpe ring. Its official name is Ring C.

The new ring proved to be relatively easy to make out, which was surprising, given its late date of discovery, and this led to the question of how it could have been missed by such diligent observers as Herschel and Schroeter. In 1796 Schroeter had gone so far as to comment that the area between Ring B and the planet appeared darker than the background sky. The Prussian astronomer Otto von Struve suggested that the ring had actually brightened since the early days, and moreover, on comparing his own measures of the dimensions of the rings with those made by earlier observers, he came to a still more startling conclusion. Ring B seemed to be collapsing inward toward the planet at an alarming rate, and he calculated that it would come into contact with the surface in about the year 2150. Struve, however, was misled by inaccuracies in the earlier measurements. There is no possibility of the dimensions of the rings having changed appreciably since Huygens's time, and the brightening of Ring C appears equally dubious. Yet though his conclusions were in error, Struve's report, which appeared in 1851, had the same effect as Schmidt's later announcement of a change in the lunar crater Linné had on lunar studies: it awakened a strong interest in a subject hitherto regarded as closed.

Soon after its discovery, the dusky inner ring was found to be semitransparent. This aspect was not easy to reconcile with the idea of solid ringlets, and in the same year that Struve's paper appeared, G. P. Bond suggested that the rings might actually be fluid. The question remained unsettled, and in 1855, Cambridge University offered as the subject of its Adams Prize essay competition the question of the extent to which the stability and appearance of Saturn's rings would be consistent with alternative opinions about their nature—whether they are "rigid or fluid, or in part aeriform, [or] consist of masses of matter not mutually coherent."

The prize was won by the sole entrant, James Clerk Maxwell, a recent graduate of Cambridge and at the time occupant of the chair in natural philosophy at Marischal College in Aberdeen, Scotland. Maxwell had been born in 1831 on a large estate at Glenlair in southeastern Scotland, and from a very early age he demonstrated a flair for mathematics—his first mathematical paper, on a method for drawing ovals, was published when he was only fifteen. Though his supreme achievement would be the creation of the theory of electromagnetism, in 1855 he was still relatively unknown, and the Saturn problem provided him with an opportunity to build his reputation.

Eventually Maxwell was able to prove that solid, fluid, and gaseous rings were alike untenable. Instead, the rings had to consist of a swarm of tiny satellites, which as seen from Saturn's great distance from the Earth blurred together into a continuous sheet. Cassini had guessed right after all.

Direct confirmation of Maxwell's calculations did not come until 1895, when the pioneer American astrophysicist James E. Keeler of the Allegheny Observatory in Pittsburgh succeeded in observing the spectrum of the rings. By means of the Doppler effect, Keeler showed that the inner edge of the rings rotated faster than the outer edge, as had to be the case if they were a swarm of satellites moving in Keplerian paths, though, tragically, Maxwell did not live to see this final triumph of his calculations— he died of stomach cancer in 1879 at the early age of forty-eight.

If the rings were indeed a swarm of satellites, what was the Cassini Division? This problem was taken up in the late 1860s by Daniel Kirkwood of Indiana University. Kirkwood had already published a theory to explain the clear zones, or "gaps," in the asteroid belt, known ever since as Kirkwood's gaps. His explanation was that they resulted from the perturbing effect of Jupiter, and in his view the Cassini Division was basically a feature of the same type. A particle in the division, he pointed out, would revolve around Saturn twice for every revolution completed by Mimas, the innermost satellite then known. This meant that Mimas would loom in the same position in the sky as seen from the particle each time it had gone around, and thus the moon's tidal forces would always be exerted at the same point in the particle's orbit. Like a person on a swing who is given a push at the same point in each cycle, the particle would acquire ever greater oscillations around its mean position. It would inevitably collide with its neighbors, in the process giving up angular momentum and moving into a new path. Thus particles in the zone would be cleared out, leaving behind a gap.

This is an example of a resonance effect, and though the 2:1 resonance with Mimas was expected to be the strongest, it was reasonable to believe that there would be weaker gaps in other positions (at 1:3, 2:5, and 3:7, for example). Later Percival Lowell, observing with his 24-inch refractor at Flagstaff, Arizona, found several new gaps in the rings whose positions agreed perfectly with resonances with Mimas, and he wrote in 1915:

"That the rings have thus been sculptured by Saturn's nearest satellite, minute as it is, subsequent investigation at this observatory has, in surprising detail, confirmed." In short, everything seemed to be falling perfectly into place, and there was no particular reason to expect that any major surprises were in store.

Everything abruptly changed with the spacecraft era. Pioneer 11 paid the first visit to the planet in August 1979 (at one point the possibility of sending it right through the Cassini Division had even been contemplated; it was never actually attempted, however, and as we shall see, this was just as well). Then came Voyager 1, which swept past the planet on November 12, 1980, and Voyager 2, which made its passage on August 25, 1981.

The Voyager cameras showed the rings in all their glory. In particular, Voyager 1's view of a crescent Saturn as the spacecraft swept behind the planet must rank among the most remarkable images of the Space Age. An enormous amount of fine structure was discovered in the rings, going far beyond anything even remotely suspected from the Earth (see fig. 10.2). It was at once obvious that there was much more to understand about the rings than had been thought. Each of the rings proved to be subdivided into thousands of narrow bands of particles, with sparser regions in between. Few of these bands are completely isolated, as true ringlets would be. But by the same token, there are few genuine gaps. Instead, the fine structure generally represents more or less smooth variations in particle density. In this book I refer to these density maxima and minima as ringlets and divisions. There is no real harm in this as long as one keeps in mind that the ringlets are not always sharply defined and that the divisions are in most cases far from completely empty. At this point, let me turn to a description of the main structures of the ring system as shown in the Voyager images, beginning with those closest to the planet and working outward (see fig. 10.3).

Ring D is a collection of sparse material lying inside the inner edge of Ring C and extending all the way to the planet's cloud tops. It was not definitely seen before the Voyagers.

Ring C, the "crêpe ring," is the innermost of the three classical rings. It extends between 74,500 and 92,000 kilometers from the center of Saturn, so its inner edge lies about 25,000 kilometers above the planet's cloud tops. Its divisions are quite regularly spaced, and in addition there are several true gaps that have eccentric ringlets lying within them. Of these, much the widest is the Maxwell Gap, located near the ring's outer edge. It contains a sharply defined ringlet about 65 kilometers wide. The inner

FIGURE 10.2. The remarkably complex structure of Saturn's rings was first revealed by the Voyager 1 spacecraft, which obtained this image from almost 1,000,000 kilometers from the planet on November 13, 1980. Sunlight is being forward scattered by the B ring, and its narrow ringlets and spoke features are clearly visible. Note that the spokes appear bright in forward-scattered light. (NASA photograph)

edge of Ring C is sharply defined, though exactly why is not known. Certainly there is no strong satellite resonance to account for it. The abrupt change in particle density between Rings C and B has not yet been satisfactorily explained.

Ring B, which extends out to 117,400 kilometers from the planet's center, is much the brightest component of the whole ring system and is a more efficient reflector of sunlight than the planet itself. There is an

FIGURE 10.3. A Voyager spacecraft view of Saturn, showing the rings from their illuminated side, which corresponds to the usual telescopic appearance as seen from the Earth. Extending outward from the planet, the main structures are Ring C, the wide Ring B (which shows some of its finer structure in this image), the Cassini Division, Ring A (with the Encke Division about four-fifths of the way out), and the slender F ring. The planet can be seen clearly even through the rings.

enormous amount of fine structure in Ring B, but unlike that in Ring C, which is highly regular, there is no obvious pattern. Also there are no isolated ringlets or true gaps in Ring B.

The Cassini Division measures 4,500 kilometers wide—about the distance from New York to San Francisco. It was formerly thought to be completely empty. On rare occasions when the rings passed in front of a star, as happened in 1917, for instance, the star was readily visible within the division, which meant that there could be little in the way of obstruction. However, it is now known that even the Cassini Division is far from empty. The density of the material in it is in fact quite as great as in Ring C, and if the plan to send Pioneer 11 through the division had ever been tried, the spacecraft would have come to grief. In another respect, too, the Cassini Division resembles Ring C: it shows the same regularity of structure. Again there are a number of discrete ringlets separated by gaps averaging 30 to 50 kilometers in width. Of these, the larger gaps near the inner and outer edges of the division contain eccentric ringlets, and that near the division's outer edge, the 430-kilometer-wide Huygens Gap, is especially interesting. It lies very near the 2:1 resonance with Mimas and is no doubt related to it.

Ring A occupies the space beyond the Cassini Division. Its inner edge lies at 122,170 kilometers, and its outer edge at 136,780 kilometers, from the planet's center. About nine-tenths of the way out on Ring A there is a prominent feature: the Encke Gap. (Actually, the name is somewhat misleading, as the sharp division here was first seen not by J. F. Encke but by James Keeler, with the 36-inch refractor of the Lick Observatory in 1888. In 1837 Encke saw merely the broad shaded region toward the middle of the ring.) The Encke Gap measures 328 kilometers across, and like the Cassini Division it contains several ringlets, of which at least two are eccentric and show a peculiar kinking and clumping. Moreover, the edges of the gap are wavy, which astronomers had concluded was due to gravitational perturbations from a small satellite orbiting within the gap. Mark R. Showalter finally identified this satellite in Voyager images in July 1990, and it has been given a name—Pan. It measures about 20 kilometers across and is embedded in a dusty ringlet that represents debris produced during impacts between the moon and small ring particles.

Ring F is located just beyond the outer edge of Ring A. It is very narrow, being only 50 kilometers wide, and thus is better described as a ringlet. Voyager 1 images show it to be made up of several strands intertwined in braidlike fashion. More about this later.

Ring G is a tenuous ring confined in the region between Mimas and two co-orbiting satellites (Janus and Epimetheus).

Ring E was officially discovered in 1980 from the Earth, though various unconfirmed reports of outer ghostly rings had been made by visual observers as early as 1907. It consists of a diffuse collection of rubble whose densest part lies just inside the orbit of Saturn's moon Enceladus.

Before Voyager it could be confidently asserted that the resonance theory adequately accounted for the various ring structures. It is now clear, however, that the rings are far more complex than had been imagined and that the resonance theory fails on many counts. Certainly it must be valid to some extent, but even where it is applicable it has had to be modified. For example, only in areas where the ring particle density is low is a clear gap produced, as predicted by Kirkwood. In denser regions, resonances set up accumulations of material known as spiral density waves rather than gaps.

One well-established resonance effect is to distort rings from their circular shape at resonance positions. At a 2:1 resonance, for example, the particles in the rings are stretched into a bilobed oval—this is the case at the edge of Ring B, where the resonance with Mimas is located. A 3:1 resonance produces a trilobed oval, and so on. As particles swing into a lobe, temporary concentrations are set up which gravitationally disturb particles lying inside. The result is a series of outwardly spiralling density waves.

A beautiful series of such waves exists in the outer part of Ring A. They are produced at resonances with the several small moons that orbit just outside the ring's outer edge. But spiral density waves are not found in the Cassini Division despite the strong 2:1 resonance with Mimas. Presumably the material there is too sparse.

The outer edge of Ring B, the Huygens Gap, and its associated ringlet all occur in the location of the 2:1 resonance with Mimas, and they must be produced by it in some way. But most of the gaps are produced by other mechanisms. For instance, the Encke Gap, as we have seen, is cleared out not by a moon orbiting outside the rings, as in classic resonance theory, but by a small moon located within the gap itself.

Indeed, small moons orbiting within or near the rings actually seem to account for most of the more exotic ring features. This is nowhere better illustrated than in the case of the peculiar F Ring. The F Ring is bounded by two small satellites, Prometheus and Pandora, both of which were unknown before the Voyager flybys. Prometheus orbits just inside the F Ring, Pandora just outside it, and the two of them act to "shepherd" the F Ring particles lying between them into a narrow "flock."

The shepherd theory was first proposed by Peter Goldreich and Scott Tremaine in 1979 to explain the set of discrete narrow rings that had been discovered two years earlier around Uranus. Goldreich and Tremaine reasoned that, left to themselves, the billions of particles making up such rings would collide with one another and soon disperse into a homogeneous sheet. But somehow the rings of Uranus were kept from spreading, and the problem was to find a mechanism to account for this.

Goldreich and Tremaine's solution was brilliant. They realized that a small satellite located just outside such a ring would travel at a slightly slower speed than the ring particles. Thus it would exert drag on particles bumped outward through mutual collisions, slowing them down and causing them to drop back into lower orbits. Conversely, an inner satellite would travel faster than the ring particles and would slightly accelerate any particles jostled inward, forcing them back out into higher orbits. In short, the satellites on either side would effectively "repel" wayward particles and gravitationally steer them back into line.

Though inspired by the slim Uranian rings, the shepherd theory was first confirmed in the case of Saturn's F Ring. Because the F Ring's shepherds, Prometheus and Pandora, are rather larger than necessary to confine the ring, they torture it into kinks, multiple braided strands, and bunched condensations. One of the shepherds even brushes the ring every seventeen years. Finally, there is a set of waves in this ring similar to those found along the edges of the Encke Gap, and though one of these is due to the gravitational disturbance of Prometheus, another points to the existence of still another small moon—so far not directly observed—about 1,180 kilometers beyond the F Ring.

Incidentally, the visible part of the F Ring, which consists of dust and centimeter-sized particles, constitutes only a small part of the total ring mass. The rest of the mass seems to reside in kilometer-sized moonlets, which are, however, only hinted at in the Voyager images.

The particles that make up the Saturnian ring system are made up of water ice together with small amounts of non-icy material, and they vary enormously in size, from tiny dust grains to blocks as huge as mountains. The particles form an ultrathin sheet in Saturn's equatorial plane— according to the Voyager data, only 100 to 150 meters thick. Their extreme thinness explains why the rings become invisible in smaller instruments when they are viewed edge-on every fourteen or fifteen years (the next such occurrence will be in 1995–96). In very large telescopes, however, they remain visible, though what is seen under these circumstances is not the extremely thin ring plane itself but so-called bending wave crests produced by Mimas. Unlike the other inner satellites, which lie in exactly equatorial orbits, Mimas has a slight inclination of 1.5°. This doesn't amount to much, but it is nevertheless significant, for it means that Mimas pulls not only outward on ring particles but also upward and downward. This produces vertical spiral density waves whose crests may reach a kilometer or more in height.

During the edge-on presentations of the rings, there are brief periods when the Earth and the Sun lie on opposite sides of the rings so that their unilluminated face is on view. Under these conditions, bright patches appear along the threadlike ring. These used to be described as knots or condensations, and some early observers actually believed that they were seeing true projections—lofty mountain peaks on the surface of the rings. It was later realized that the condensations appear in the exact positions of Ring C and the Cassini Division, and they are due to the fact that while sunlight is so completely back-scattered by Rings A and B that none of it gets through, the sparser material of Ring C and the Cassini Division allows some forward-scattering, so these areas appear luminous. The view under these conditions is exactly analogous to those obtained by Voyager from behind the planet, when the rings were seen by forward-scattered sunlight (see fig. 10.4).

Finally, mention should be made of the radial bands or "spokes" observed by the Voyagers on the surface of Ring B (see fig. 10.2). Though in 1977 one keen-eyed visual observer, Stephen J. O'Meara, had sketched such features using a 9-inch refractor, no one had paid any attention. Indeed, radial structure in the rings was totally unexpected, and there is a good reason for this. Remember, the ring particles exhibit differential rotation. Those at Ring B's inner edge travel in Keplerian orbits having periods of 7.9 hours, while those at the outer edge have periods of 11.4 hours. This means that even if a radial feature developed, it ought to be immediately disrupted by shear. Yet radial features as much as 10,000 kilometers in length do exist. At first, astronomers were completely baffled as to what they might be, but the Voyager images showed beyond doubt that they rotate along with the magnetic field of Saturn, whose period of 10 hours, 39.4 minutes is that of the planet's core. This led to the identification of the spokes with electrostatically charged particles levitated out of the ring plane by Saturn's magnetic field and then swept along by the magnetic field lines. They are most apt to appear when a specific magnetic longitude is roughly aligned with Saturn's morning terminator, and since the Voyager results became known there have even been a number of reports of sightings with modest instruments.

How did the rings come to be? The first serious attempt to answer this question came in 1848, when the French mathematician Édouard Roche calculated the distance at which tidal forces acting on a liquid satellite would exceed the binding force holding it together. This distance is called Roche's limit, and for Saturn it is equal to 2.44 times the radius of the planet. Saturn's rings lie well within the limit. Thus, according to Roche, the rings are the ruins of a hapless moon that long ago veered too close to the planet and was ruthlessly torn to pieces.

For the sake of simplifying the calculations, Roche assumed that his satellite was fluid. When the same problem was later reworked assuming a solid satellite, it was found that a small solid satellite would not break apart no matter how close it came to the planet, whereas a large one would do so, but only if it approached well within the zone occupied by

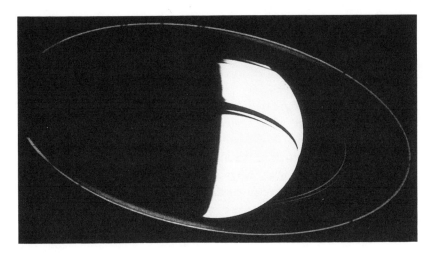

FIGURE 10.4. This Voyager spacecraft view shows the rings from their unilluminated side. In several places, including the Cassini Division, sunlight sifts through the rings, but the densest part of Ring B is visible only because of "Saturnshine," sunlight reflected from the ball of the planet. The braided structure of the F ring makes it appear discontinuous here. (NASA photograph)

the current rings. For this reason, Roche's original theory had to be abandoned, and it is now believed that rings form when small satellites are smashed by comet impacts. If the shattered satellite lay outside Roche's limit, its remnants might reassemble into a moon, but inside Roche's limit this is impossible, and the fragments continue to orbit the planet in the form of rings.

Saturn's rings appear to consist of remnants of several different catastrophes. Of the main rings, Ring C appears to be the most ancient, though it is probably less than 100 million years old. Were it much older, it would be darker than it is, due to soiling of its icy particles by micrometeoroids. Judging from the proximity of small moons to its outer edge, Ring A must have formed less than 20 million years ago. Thus the rings we see are relative newcomers to the Saturnian scene, and though they will last for tens of millions of years yet, they are far from eternal. Mutual collisions and micrometeoroid impacts will gradually grind them to dust, which is then lost from the system. The spiral density waves described earlier involve a net transfer of angular momentum from the ring particles to the small moons lying just outside Ring A, with the result that these moons are retreating away from the planet even as the rings are collapsing inward. Thus Struve's idea of collapsing rings proves not to have been so farfetched. He was simply wrong about the time scale, which instead of being a few centuries is actually on the order of about 100 million years.

Saturn's atmosphere is partly driven as Jupiter's is, by an internal heat engine, and its jet currents have even greater velocities, though there are fewer of them. The westerly Equatorial Current is the fastest stream, with wind speeds reaching 1,800 km/hr—two-thirds of the speed of sound there.

The Saturnian clouds as seen in small telescopes are cold and marbled in appearance, and the bands into which they are drawn are less distinct than Jupiter's. The reason for this is that the cloud forms can only be viewed through a high, dense aerosol haze. Spots are also less frequent, though they do occur—for example, a one-third-sized version of Jupiter's Great Red Spot was discovered by Voyager 1 in high southern latitudes.

Then there are the spectacular white spots, which have been recorded once every thirty years. The first on record was Asaph Hall's Spot, which appeared in the Equatorial Zone in 1876. This spot played an indirect role in leading to the discovery of the Martian satellites, as from his observations Hall was able to work out a new rotation period for Saturn of 10 hours, 14 minutes, correcting what had hitherto been accepted as the standard by sixteen minutes. It was this experience, he later recalled, that made him skeptical of such statements in textbooks as "Mars has no moons."

A white spot was observed in the northern latitudes by E. E. Barnard with the Yerkes 40-inch refractor in 1903. Then in 1933 an especially brilliant spot appeared in Saturn's Equatorial Zone. It was first recorded by W. T. ("Will") Hay, a well-known British stage and screen comedian, whose private observatory at Norbury, outside London, was equipped with a 6-inch refractor. The spot was so brilliant however, that there were a host of independent discoverers. As with previous white spots, Hay's rapidly lengthened toward the east over the next few weeks. By the time it finally broke up, it stretched halfway around the planet.

In 1960 another white spot, discovered by the South African astronomer J. Botham, appeared in Saturn's far northern latitudes. A thirty-year cycle was thus established that corresponded to the interval between the times when Saturn's north pole was tilted maximally toward the Sun. This condition applied again between December 1989 and February 1990, and not long afterward another white spot appeared. It was discovered on September 24, 1990, by an amateur astronomer, Stuart Wilber, who used a 10-inch reflector at Las Cruces, New Mexico. I first saw it several days later, when it was a small, intensely brilliant cloud, easily visible with only a 2-inch refractor. Within a week it had been spread by the strong equatorial winds into an oval cloud 15,000 kilometers long. The area of disturbance continued to elongate, and by late October whitish clouds encircled the whole globe. They were separated by a dark band along the line where they were being sheared apart by the somewhat faster currents just north and south of the equator. The beautiful festoons along the northern edge of the whitish clouds were captured in images obtained by the Hubble Space Telescope in early November 1990 (see fig. 10.5).

Somewhat reminiscent of what occurs during disturbances of the South Equatorial Belt of Jupiter, Saturn's white spots develop when a bubble of sublimating ammonia, originating in a distinct "hot spot" on the planet, rises to an altitude of perhaps 250 kilometers above the usual cloud tops. As it gives up its latent heat, the ammonia chills to form brilliant, billowy clouds of ammonia-ice crystals.

Saturn's deep structure is similar to Jupiter's. Both planets are made up chiefly of hydrogen and helium, though on Saturn there is relatively less helium. Also, because Saturn is less than a third as massive as Jupiter, the transition from liquid to metallic hydrogen occurs at a deeper level within the planet, about halfway from the cloud tops to the center. As on Jupiter, the rotation of the metallic hydrogen core generates a powerful magnetic field. Saturn's magnetic field has only one-twentieth the intensity of Jupiter's, but it is still a thousand times more powerful than the Earth's, and it is unique in lining up almost perfectly with the planet's axis of rotation, so a compass needle there would point due north. Finally, at the very center of the planet there is thought to be a small rocky core about equal in diameter to the Earth.

Saturn has a fascinating bevy of moons—in number not exceeded even by mighty Jupiter itself. Counting Pan, the small moon recently identified within the Encke Gap, eighteen are so far known. Of these, twelve were discovered from the Earth and six by the Voyager spacecraft.

The largest, Titan, was discovered by Christiaan Huygens on March 25, 1655. It circles Saturn once every 16 days at a distance of 1,221,000 kilometers from the planet's center. At 5,150 kilometers, Titan's dimensions are planetary—it is larger than Mercury and only a little smaller than Jupiter's Ganymede. Moreover, Titan has its own atmosphere, the existence of which was suspected as early as 1908 by the Spanish astronomer J. Comas Solá from the pronounced limb darkening of its orangish

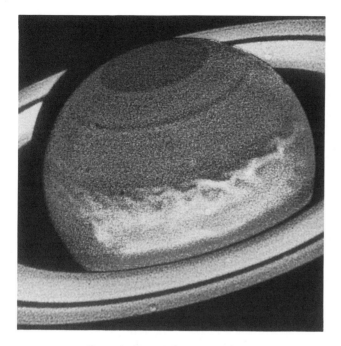

FIGURE 10.5. Saturn's Great White Spot of 1990, which appeared as a small, brilliant cloud in late September, when it was discovered by amateur astronomer Stuart Wilber of Las Cruces, New Mexico. By the time the Hubble Space Telescope obtained this image, on November 9, the whitish clouds had spread around the whole globe. This was by far the most spectacular eruption on Saturn since 1933. The moonlike image in the bottom center is the result of a defect in the photograph. (NASA photograph)

brown disk. Its existence was not confirmed, however, until the end of World War II, when G. P. Kuiper detected methane with a spectroscope.

The atmosphere of Titan is 95 percent nitrogen, with a trace of methane and other hydrocarbons, and the surface pressure is half again as great as the Earth's. It is an inefficient heat trap, however, and the temperature on the surface is a chill −180°C. The hydrocarbons form a dense orangish brown aerosol "smog," thus accounting for the distinctive color of its disk.

The smog is hopelessly opaque, and as a result the Voyagers could make out nothing whatever of the surface. They did manage, however, to obtain some attractive images from the satellite's night side that show the same prolongation of the cusps into a ring as had long been familiar from Venus near inferior conjunction—this is, of course, simply an effect of atmospheric scattering of sunlight. The surface is believed to be covered at least in part by oceans of liquid methane, though recent radar observations indicate that there are also land masses of some sort.

Following Huygens's discovery of Titan, G. D. Cassini, in 1671–72, added two more moons. Huygens had given no name to his satellite other than "Saturn's moon," but as noted in chapter 2, Cassini now proposed the name "Louisian stars" after Louis XIV, his patron and the founder of the Paris Observatory, of which Cassini was the first director. Cassini called to the attention of the Sun King that his discoveries had raised the number of known bodies in the Solar System to precisely fourteen, something which no doubt gratified the royal ego. Unfortunately, the Solar System was not constructed with the number fourteen in mind; Cassini himself found two additional satellites in 1684.

Needless to say, "Louisian stars" proved no more popular than Galileo's "Medicean stars" had been for the satellites of Jupiter, and William Herschel, who added two more moons in 1789, simply adopted numbers. It was his son Sir John who proposed the mythological names for the first seven which are still in use today.

The most interesting of Cassini's finds, Iapetus, is 1,440 kilometers across and is located at a distance of 3,560,000 kilometers from Saturn. It is remarkable for being nearly two magnitudes brighter when west of Saturn than when east. Cassini himself suggested that one hemisphere must be a better reflector than the other. This was the only reasonable explanation, and it is now fully confirmed. The Voyager images show a broad black swath extending 220° of longitude and 110° of latitude along the satellite's leading hemisphere (the side facing in the direction of the tidally locked satellite's motion in its orbit). It appears to be some sort of dark material deposited on an underlying icy surface. At one point it was suggested that the source of this material may have been the small outer satellite Phoebe, whose surface is also dark. However, closer scrutiny revealed that Phoebe's color is black, while the dark covering of Iapetus has a slight reddish tinge, so this idea had to be given up. It is now thought that the impact of a comet provided the source of the dark material. Such an impact would have started Iapetus rocking about its polar axis, and as carbon compounds scattered from the impact were swept up by the satellite as it traveled in its orbit, they would have been strewn on the surface in the observed distribution.

Cassini's other satellites, Tethys, Dione, and Rhea, lie closer to Saturn than Titan, at distances ranging between 294,700 and 527,200 kilometers. Tethys is 1,530 kilometers across, and the other two are rather more than 1,000. It is certainly a credit to Cassini's skill that he was able to locate them using the unwieldy aerial telescopes of his day.

The dominant feature on Tethys is a large fracture, Ithaca Chasma, which stretches all the way from one pole to the other. Also notable is the 400-kilometer-wide crater Odysseus. Dione (fig. 10.6) shows an intricate system of bright wisps believed to consist of surface frost. The wisps converge on a craterlike feature, Amata. One hemisphere of Dione is smooth and dark; the other is brighter and contains numerous impact craters, of which the two largest have been named Aeneas and Dido. The difference in reflectivity of the two hemispheres approaches that found on Iapetus, though on Dione it is the trailing hemisphere that is darker.

Rhea too has a smooth and dark trailing hemisphere and a heavily cratered leading hemisphere. The largest crater, Izanagi, is 200 kilometers wide and occupies a position not far from the boundary between the two zones.

We come next to the two satellites found by William Herschel in 1789, Mimas and Enceladus, which lie at distances of 185,000 and 238,100 kilometers from Saturn, respectively. Mimas is 390 kilometers in diameter and shows a battered face, dominated by a ringed basin 130 kilometers wide—Herschel. The impact that formed this feature must have come perilously close to shattering the moon. Enceladus, 500 kilometers wide, has a surface that is smooth and youthful looking and that has evidently undergone major reworking. In particular there is an area of ridged plains that is suggestive of crustal movements. Enceladus completes each orbit in 1 day, 8 hours, 53 minutes and thus lies at a 2:1 resonance position with Dione, whose orbital period is 2 days, 17 hours, 41 minutes. This means that Enceladus is subject to strong tidal forces from more massive Dione, and as with Jupiter's Io, which is in the same sort of resonance lock with Europa, tidal flexing and bending keeps the interior warm, allowing continuing geological activity to take place.

Incidentally, all of the satellites mentioned so far are within reach of amateurs, though Enceladus requires a 10-inch telescope and Mimas is even more difficult, being swamped in glare because of its nearness to Saturn. I must admit that my 12 1/2-inch reflector seldom shows it.

There is one other Saturnian satellite of possible interest to amateurs—Hyperion. Discovered by G. P. Bond at Harvard in 1848, it orbits between Titan and Iapetus, and at magnitude 14 it can just be captured in a 10-inch

FIGURE 10.6. Dione, as photographed by Voyager 1 from a distance of 240,000 kilometers. This view shows the hemisphere that faces Saturn. The large crater that dominates the heavily cratered region to the left of center is Aeneas, which is about 100 kilometers across, while the bright, wispy features at the extreme right originate from Amata, a feature in the other hemisphere. (NASA photograph)

reflector around its greatest elongations. It has an irregular shape, measuring 350 kilometers along one axis but only 200 kilometers along another, and there can be little doubt that it is a fragment of some shattering impact. As a result of its irregular shape and the disturbance of its orbit by Titan, Hyperion tumbles wildly as it travels around Saturn. Its rotation is actually "chaotic," that is, it changes without any predictable pattern.

Even more remote from Saturn is Phoebe, discovered photographically in 1898 by W. H. Pickering but unfortunately not well imaged by either Voyager. The most interesting thing about it is its orbit, which is retrograde and tilted to that of Saturn by an angle of 150°. Also, though its period of revolution is 550 days, Phoebe rotates in only 9 hours. It would not be at all surprising if it eventually proved to be a captured asteroid.

We come, finally, to the smaller moons, which are interesting in their own right. Several occupy Lagrangian points in orbits with other moons.

Helene, which was discovered by P. Lacques and J. Lecacheux of the Pic du Midi Observatory when the rings were edgewise in 1980, shares an orbit with Dione, while two Voyager finds, Telesto and Calypso, are co-orbital with Tethys.

Then there are the strange moons Janus and Epimetheus. Janus measures 220 by 160 kilometers and Epimetheus 140 by 100 kilometers. They are both located at about 151,400 kilometers from Saturn, and they have nearly identical periods of right around 16 hours, 40 minutes. The distance between them is less than the sum of their diameters, but because their periods differ by about half a minute, they encounter one another once every four years (the last time was in January 1990). When they do, they swap orbits, with the one that had been slightly closer to Saturn moving farther out and vice versa. Recent studies indicate that the densities of these moons are actually less than that of pure ice. This means that they must be highly porous, and they are probably no more than clumps of icy ring fragments loosely bound together by gravitation.

Janus and Epimetheus are the outermost of five small satellites that obviously formed in close association with the rings and that are intricately involved with their structure, so they are aptly described as ring-moons. The others are Atlas, which lies just outside the A Ring and which seems to be responsible for defining its sharp boundary, and just beyond it Prometheus and Pandora, already familiar as the shepherds of the F Ring. There can be little doubt that these ringmoons are fragments from the impact that produced the A Ring itself.

Were it not for the anachronism, one might almost imagine that the poet Henry Vaughn had Saturn in mind when he wrote:

> I saw Eternity the other night
> Like a great ring of endless light,
> All calm as it was bright.

But though they may seem to be emblems of eternity, the rings are not eternal; they have their own cycles of growth and decay. As we have seen, even now they are in the process of weathering away, but they can be replenished as well. There must be many blocks a kilometer or more across within the ring system itself, and the ringmoons as well contain the seeds of future glories, needing only a random comet strike to create new spectacles of rings.

CHAPTER ELEVEN

Uranus

William Herschel, the discoverer of the seventh planet of the Solar System, was born in Hanover in 1738 of a musical family. As a young man he served as a bandsman in the Hanoverian Guard, but the military life did not appeal to him, and after coming under fire at the battle of Hastenbeck in 1757, he left for England, which was then ruled by a Hanoverian, George II. Herschel was variously engaged as a copyist, teacher, performer, and composer before finally being hired as an organist in the Octagon Chapel in the brilliant resort town of Bath, a position that gave him a good income. This he supplemented by giving music lessons as well as additional concerts—aside from the organ, he was skilled in the violin, oboe, and harpsichord. Meanwhile, he had begun to take a serious interest in astronomy, and in the early 1770s, soon after he had been joined in England by his younger sister Caroline, he wrote:

> The great run of business, far from lessening my attachment to study, increased it, so that many times after a fatiguing day of 14 to 16 hrs spent in my vocation, I retired at night with the greatest avidity to unbend the mind (if it may be so called) with a few propositions in Maclaurin's *Fluxions* or other books of that sort.
>
> Among other mathematical subjects, Optics and Astronomy came in turn, and when I read of the many charming discoveries that had been made by means of the Telescope, I was so delighted with the subject that I wished to see the heavens and Planets with my eyes thro' one of those instruments.

FIGURE 11.1. William Herschel (1738–1822) (Portrait by John Russell, 1794)

The telescopes available in Herschel's day were far from satisfactory, so he decided to build his own. He settled on the reflector type and soon became such a diligent grinder of mirrors that at times he worked for sixteen hours straight. In addition to reading to him, Caroline was assigned the task of putting the food "by bits into his mouth."

Herschel's first successful mirror, of 4 1/2-inch diameter, was finished only after many failures, but he learned from his mistakes, and eventually he was able to equip himself with a 6 1/2-inch reflector that was better than any of the instruments then in use at the Royal Observatory at Greenwich. Herschel was, moreover, as diligent a user of his telescopes as he was a builder of them, and it was not unknown for him to dash from the harpsichord between acts at the theater in order to observe.

Herschel was not one to accept anything on faith, and during his lifetime he completed several telescopic "reviews" of the whole heavens. Between 10 and 11 o'clock on the evening of March 13, 1781, he was in the midst of one of these reviews, using his 6 1/2-inch telescope in the garden of his house at 19 New King Street. In the course of these observations he had come to the interesting region between the Crab Nebula and the open cluster Messier 35. There he noticed what one observer in a million would have noticed: "In the quartile near ζ [zeta] Tauri the lowest of two [stars] is a curious either nebulous star or perhaps a comet." On the next clear night, March 17, Herschel "looked for the comet or nebulous star, and found it is a comet, for it has changed its place." If so, it was a comet as yet too far from the Sun to have developed a tail, but there was nothing particularly unusual about that, and at once Herschel wrote a brief paper, "An Account of a Comet," which he presented to the Bath Literary and Philosophical Society. Continuing his observations, Herschel on March 28 jotted down in his log: "The diameter is certainly increased, from which we may conclude that the Comet approaches us." In this he was mistaken. We now know that the diameter of the disk actually decreased slightly over the period covered by his observations. Here Herschel's preconceptions as to what he had found led him astray.

Meanwhile, other observers had been informed of the discovery and had begun tracking Herschel's "comet." One of them, Nevil Maskelyne, the Astronomer Royal at Greenwich, had worked out a provisional orbit by late April and had reached the surprising conclusion that what Herschel had found was "as likely to be a regular planet moving in an orbit very circular around the Sun as a comet moving in a very eccentric ellipsis." Moreover, the object had still failed to develop a nebulous appearance or a tail, and Charles Messier, the best-known comet discoverer of the day, congratulated "Monsieur Hertsthel at Bath" on the discovery, adding that "nothing could be more difficult than to recognize it, and I cannot conceive how you were able to return several times to this star—or comet . . . since it had none of the characteristics of a comet."

By midsummer 1781, all the evidence had begun to point to the conclusion that Herschel had indeed discovered a new planet, the first in modern times—though some astronomers, Herschel included, hesitated to call it one for a few more months. As the orbit was more accurately determined, the conclusion became inescapable. The planet orbits the Sun at a mean distance of 2,387 billion kilometers, so it is almost twice as far from the Sun as is Saturn, and it requires 84 Earth years to complete each orbit.

Naturally, Herschel immediately became famous. King George III wanted to have a look through his telescope, and on July 2, 1782, Herschel came to Windsor to show His and Her Majesties Jupiter and Saturn. The King was obviously impressed, for soon afterward he granted Herschel a generous pension that allowed him to give up music and pursue astronomy full-time. The only requirement was that he take up residence near Windsor. Thus he came first to Datchet, then to Clay Hall, Windsor, and finally in April 1786 to Observatory House, Slough, in the garden of which he set up what was for a long time the largest telescope ever made, his mammoth 40-foot reflector with its 4-foot mirror. It was not finally surpassed in size until 1845, when William Parsons, the third earl of Rosse, completed his 6-foot reflector at Birr Castle in Ireland.

Not long after King George granted Herschel his pension, the astronomer reciprocated by formally proposing for the planet the name of Georgium Sidus—the Star of George. "I cannot but wish to take this opportunity," he wrote, "of expressing my gratitude, by giving the name ... to a star, which (with respect to us) first began to shine under His Auspicious Reign." Herschel's flattery was no less fulsome than Galileo's or Cassini's had been. Later, George became completely insane (one of his delusions was of seeing Hanover from Windsor Castle with one of Herschel's telescopes). Nevertheless, the *Nautical Almanac* stubbornly clung to "the Georgian" through the edition of 1851. The name was never adopted on the Continent. There, "Herschel" was popular for a time, while other names were also proposed. The name that eventually prevailed, Uranus, was suggested by Johann Bode.

Following Herschel's discovery, astronomers found that Uranus had been recorded on a number of earlier occasions as an ordinary star. England's first Astronomer Royal, Rev. John Flamsteed, for example, had recorded it on December 23, 1690, while making observations for his star catalog. He entered it as an ordinary star, 34 Tauri, and made six more observations of it between 1712 and 1715, including three in the course of a single week. It was also seen by Germany's Tobias Mayer in 1756, but a French observer, Pierre Charles Lemonnier, holds the record. He saw it twelve times in all, including six times in nine days in 1769, without recognizing it as a planet—and lived to rue the fact. Lemonnier seems to have been a rather disorganized man (one of his Uranus observations was later found scrawled on an old bag that had contained hair perfume). Yet in fairness to him it should be pointed out that his observations of 1769 were made at a time when Uranus was moving slowly near opposition, so changes in its position would have been less than obvious.

Clearly Herschel's discovery does great credit to his telescope and his careful methods, yet in fact Uranus is not a difficult object to observe. In a dark sky it can even be made out with the naked eye, appearing as a sixth-magnitude "star." But its remoteness has made it difficult to scrutinize in any detail from the Earth. Its aquamarine disk, 4 seconds of arc in diameter, can be made out in modest instruments, but little else can be seen.

By 1829 observers had established that Uranus has an unusual axis of rotation. The axis is tilted by some 98° from the perpendicular, so the planet appears to be rolling around its orbit on its side. This means that our view is sometimes over the equator, as in 1924 and 1966, and at other times over one or the other pole, as in 1945 and 1987. It goes without saying that the seasons on Uranus are highly unusual. For twenty-one years, parts of one Uranian hemisphere are submerged in night, then the Sun rises and brings on an equally monotonous "day."

Vague markings on the planet have been reported from time to time by telescopic observers—usually, by analogy with those of Jupiter and Saturn, dusky belts and lighter zones. Thus E. M. Antoniadi recorded dark equatorial belts with the great Meudon refractor in 1924, while a few years later, in 1936, he drew only a dark polar cap. Antoniadi was a credible witness and was using one of the largest telescopes in the world, but what is one to make of the sightings of the Scottish observer C. Roberts? Using only a 6 1/2-inch reflector in 1896, he found the belts "so persistently visible I ventured to draw them, notwithstanding the fact that much larger telescopes have shown nothing." I hasten to add that Roberts was also a virtuoso observer of the "canals" of Mars—so much so that Antoniadi, who in 1896 had been director of the Mars Section of the British Astronomical Association, decided that it was probably safest to exclude Roberts's observations from the section's map of that planet rather than to "overcrowd our already crowded chart with the most daedalian canal network ever devised."

As we shall see, these impressions of belts on the planet's surface must now be regarded with skepticism, but it is easy to see how visual observers could have imagined such things. Uranus, after all, is clearly a giant world like Jupiter and Saturn—its diameter is 51,118 kilometers through the equator—and the power of suggestion seems to have done the rest.

As it is a giant world, there was never the slightest doubt that Uranus was made up mostly of gas, and hydrogen and helium were bound to be the leading constituents, but the spectroscope showed that methane (CH_4) and acetylene (C_2H_2) were also present. On Uranus, where the temperature at the cloud tops (about −208°C) is colder than on Jupiter and Saturn, methane is able to freeze out. Thus a layer of methane clouds overlies the ammonia clouds that form the next highest layer. Methane absorbs reddish light but lets bluish light through, whence the planet's distinctive aquamarine color.

In the 1930s J. H. Moore and D. H. Menzel used the Doppler shifting of the lines in Uranus's spectrum to determine the rotation period. Their result, 10 hours, 49 minutes, was long quoted in textbooks, but it was later found to be widely in error. By the 1970s the figure had risen to a range of 15 to 17 hours. On the whole, however, very little was known about the planet, and even the planet's fascinating system of rings was not discovered until 1977.

Oddly enough, on several occasions between 1787 and 1794, William Herschel himself received the impression that a ring was present, which naturally raises the question of whether he may have glimpsed one of them. After all, he used a large telescope, and the English skies in the eighteenth century were still very dark and unpolluted. The answer, unfortunately, is no. The rings are far too dark for Herschel to have seen anything of them, and his sightings have now been satisfactorily explained as due to ghost reflections produced by the "front-view" optical system he used in his larger reflectors.

The actual discovery of the rings did not come until March 10, 1977, when James Eliot, Edward Dunham, and Douglas Mink were observing Uranus's occultation of the ninth-magnitude star SAO 158687 in order to obtain a more reliable value for the planet's diameter. They succeeded in this, but they also found something entirely unexpected. Just before the star passed behind the planet, five unexpected minima in the intensity of the light appeared in the recording, and the same minima were repeated on the following side. They realized at once that the only possible explanation was that Uranus was surrounded by at least five exquisitely narrow rings, which have been given the Greek letter designations α, β, γ, δ, and ϵ.

More sensitive ground-based observations later raised the number of rings to nine, while Voyager 2 has added a tenth. The rings lie between 42,000 and 51,000 kilometers from the planet's center, and each is a narrow, discrete "ringlet" only a few kilometers wide—the narrowest measures less than a kilometer. I shall have more to say about them presently.

The two-hundredth anniversary of Herschel's discovery of Uranus came on March 13, 1981. Two centuries of Earth-based observation and study had yielded disappointingly little. I remember peering at Uranus with the 24-inch Clark refractor at Lowell Observatory in July 1982. The disk was a beautiful pale aquamarine, but even in this superb instrument it was absolutely featureless. Viewing Uranus pole-on, the four brightest satellites were ranged around the planet like a miniature Solar System. I could not help feeling that these cold and distant worlds, which seemed to have some of the unapproachable dignity of Greek sculpture, formed a system that might well repay closer acquaintance, but what they would then show it was impossible even to guess.

Then came Voyager 2, which had been slingshotted toward Uranus after its 1981 encounter with Saturn. In the closing hours of that flyby, however, the spacecraft's camera platform had jammed, causing the loss of the best images of Tethys and Enceladus. Fortunately, the problem could be corrected, and the trip out to Uranus passed without further incident.

Since the goal of the spacecraft was to continue on to Neptune after the Uranus encounter, Voyager 2 was targeted to sweep through the Uranian system at a precise distance from the planet's center, but in December 1985—only a few weeks before the scheduled flyby—it was found that the planet's satellites kept showing up in the wrong positions. The explanation was that the planet was slightly more massive than had been thought. This also meant that a course correction was required to change the spacecraft's trajectory by several hundred kilometers. Otherwise it would have missed its next target, Neptune, by millions of kilometers. With these adjustments, the spacecraft came in right on the mark. Its closest approach to Uranus, on January 24, 1986, was 81,543 kilometers above the cloud tops, or only 15 kilometers from the aim point.

In the spacecraft's approach images, the planet's disk appeared completely blank, the deeper cloud features being hidden by a thick haze that shrouds the sunlit pole to a latitude of about 35°s. This haze is made up of hydrocarbon droplets or crystals synthesized from acetylene warmed by the dim Uranian Sun. Using special electronic enhancement techniques, however, it was possible to bring out a few spots. One, at 27°s, had a rotation period of 16.9 hours, and another at 40°s had a period of 16.0 hours. (Needless to say, none of these spots would have been even remotely within reach of visual observers.)

The rotation periods mentioned above are those of the atmosphere. The next question, obviously, is how they compare with the "deep" rotation of the planet, for once we know this, the atmosphere's circulation pattern can be worked out. On Jupiter and Saturn the deep rotation periods had been discovered by measuring radio bursts from particles trapped in their magnetic fields, but before Voyager 2 it was not even known whether Uranus had a magnetic field.

Uranus finally broke its radio silence only two days before Voyager's closest encounter, and it was found to have a magnetic field with an intensity fifty times that of the Earth but tilted by an extreme angle (almost 60°) to the planet's axis of rotation. Were the same situation true for the Earth, compass needles would point toward Florida. To make matters still more confusing, the south pole of rotation, which at present points toward the Sun, actually lies nearer to the north magnetic pole. The bizarre tilt means that as the planet turns, its magnetic field undergoes tortuous gyrations. Voyager's radio observations indicated that the interior rotates with a period of 17.24 hours, rather slower than the periods observed for atmospheric features.

Unlike Jupiter, Saturn, and Neptune, Uranus lacks an internal heat source. As a result, solar energy alone must drive its circulation. Because the planet is for all practical purposes tipped onto its side, however, its poles receive more heat than its equator during the course of a Uranian year. Thus, if the planet did not rotate at all, one would expect the atmospheric circulation to consist of a simple north-south current converging on the south pole, with a lower-lying return current. The rotation deflects the winds into zonal jets like those found on the other giants, and these are so effective at redistributing heat around the planet that temperatures at the poles and the equator and on the day and night side all lie within only 2° of a uniform $-221°C$.

Unlike Jupiter and Saturn, which have strong equatorial westerlies, Uranus has an easterly equatorial jet. The wind speed is about 350 km/hr. At higher latitudes, the flow direction changes and wind speeds increase.

Uranus is obviously a world of many crotchets, and at one point an "act of God" explanation seemed to be the most logical way of explaining all of them. The planet, so it seemed, had been the victim of a glancing impact early in its history which toppled it over onto its side, resetting the orientation of its axis of rotation so that it no longer approximated that of the magnetic field. Not only that but such an impact would have produced stirring only in the planet's outer layers, resulting in less convection in the interior than was the case for the other gas giants. The heat produced during the planet's formation would thus have remained trapped rather than leaking out, as on Jupiter, Saturn, and Neptune.

All this, I repeat, seemed extremely convincing—until Voyager arrived at Neptune and showed that, despite having a normally tilted axis of rotation, it too has a cockeyed magnetic field. Thus the impact theory has lost much of its allure. Instead, it is now believed that Uranus and Neptune have a drastically different internal structure from the other planets. On planets where the magnetic fields are normally oriented, the fields are thought to be generated by convection within a molten metallic core. Uranus and Neptune, on the other hand, resemble stars known as pulsars, which behave magnetically as "oblique rotators." Their skewed magnetic fields are thought to be produced through the convection of an electrically conducting fluid lying outside a presumably solid core. In the case of Uranus and Neptune, this electrically conducting fluid is probably liquid water.

Uranus has a fascinating system of moons, of which five were known before Voyager 2. The two largest were discovered by William Herschel on January 11, 1787, with his 18.7-inch reflector. They were not seen by anyone else until 1828, when Herschel's son Sir John made them out. It was Sir John, by the way, who proposed their names, Titania and Oberon, after the king and queen of the fairies in Shakespeare's *Midsummer Night's Dream*.

William Herschel actually believed that he had found no less than six Uranian moons, and only in 1850 did the true situation become clear. William Lassell, a brewer who set up a powerful 24-inch reflector at his home near Liverpool, proved that there were only two moons in addition to Titania and Oberon. They were named, again by Sir John, Umbriel and Ariel after two of the sylphs and gnomes in Alexander Pope's *Rape of the Lock*. It is just possible that William Herschel himself had glimpsed Ariel on one or two occasions, but the rest of his suspect satellites can only have been faint stars in the same field with Uranus.

Certainly Umbriel and Ariel are much more difficult objects to make out than Titania and Oberon. Not only are they fainter, they also lie closer to the glare of the planet. In dark skies I have succeeded in making out Titania and Oberon with my 12 1/2-inch reflector, but the other two require much larger instruments and are essentially beyond the reach of amateurs. All four lie between only 12 and 44 seconds of arc of the planet's small disk.

Lassell searched for other moons without success, and indeed there were no further developments until 1948, when G. P. Kuiper, working photographically with the 82-inch reflector at McDonald Observatory in Texas, added a fifth satellite: Miranda. It lies only 130,000 kilometers from Uranus, 83,000 kilometers beyond the ε ring. From the Earth it registers as nothing more than a tiny speck of light in long-exposure photographs, and it has never been seen visually.

It was naturally assumed that, because of their enormous distance from the Sun, the Uranian moons would be made up more of ice than of rock, and this is indeed the case. All are extremely lightweight. Even Titania and Oberon have masses of only one-eighth to one-ninth that of the Earth's moon.

Imaging the surfaces of the moons posed a difficult challenge for Voyager

2. The dim lighting conditions at Uranus meant that relatively long exposures were required, but because the spacecraft swept past the satellites at high speed, this would have resulted in images that were useless blurs. Fortunately, the Voyager engineers hit on the solution of firing the spacecraft's "attitude-control" thrusters so as to turn the entire spacecraft at the proper rate in order to keep the object of interest in a fixed position in the camera's field of view—in effect, "panning" the camera.

The results were spectacular beyond all expectation. In the Voyager 2 images, Oberon—the outermost and next to largest of the satellites at 1,545 kilometers across—shows a primitive, heavily cratered surface. Several of the craters have dark floors, suggesting they have been filled by some kind of lava. Titania, the largest satellite, measures 1,584 kilometers across (about half the size of the Earth's moon) and is also heavily cratered, but the most arresting features are its rifts. The largest, Messina Chasmata, runs for hundreds of kilometers. The rifts are believed to have formed when Titania cooled. The outer layers of the satellite would, of course, have cooled first, trapping liquid water below the surface, and with further cooling the subsurface water would have frozen into ice. Since ice occupies more volume than liquid water, this would have produced an upward expansion, straining and cracking the overlying crust. The same process probably produced the cracklike features on the surface of Jupiter's moon Europa.

In approaching the planet, we come next to Umbriel and Ariel, both of which have diameters right around 1,200 kilometers. Umbriel's surface looks ancient and is surprisingly dark, so in every way it merits the name given it by Sir John Herschel from Pope's *Rape of the Lock*:

> Umbriel, a dusky melancholy Spright,
> As ever sully'd the fair face of Light.

Though similar to Uranus's other moons in being rich in water ice, Umbriel's surface was coated at some point with sooty material. Even its impact craters are dark and, except for a single bright ring in a large crater, there are none of the gleaming ejecta blankets found on other icy moons.

Ariel at first looked rather like Titania in that it too has a system of large rifts. But on Ariel the rifts give rise to a far more extensive network of branching, smooth-floored valleys, which cover the entire visible hemisphere. Also, it has fewer impact craters than Titania. Perhaps a closer analogy is Saturn's Enceladus, where there is a somewhat similar system of rifts and where the craterous terrain has largely been obliterated by ice floes. However, whereas on Enceladus this resurfacing testifies to a warm interior reasonably explained by the moon's being in resonance with Dione, Ariel is not currently in resonance with any other moon.

Another curious feature is that Ariel's valleys are filled with material that has a smooth convex shape, which suggests that it was extruded as ice rather than as a liquid (which would, of course, level out to produce a flat surface). Under the frigid conditions of Ariel, it is difficult to understand how water ice could possibly "bend" into such shapes at all, considering that ice there would be quite as hard as rock. It seems likely that the process involved more exotic ices containing, in addition to water, small amounts of methane, ammonia, carbon monoxide or nitrogen.

We come, finally, to Miranda (fig. 11.2)—that mere speck on Earth-based photographs but as revealed by Voyager 2's cameras a remarkable world sure to appear on anyone's list of the Seven Wonders of the Solar System. Luckily, Voyager got some very good views indeed, approaching to within 28,000 kilometers of Miranda, closer than to any of the other satellites, and capturing features as small as a few hundred meters across—the best resolution for any object imaged by the spacecraft.

With a diameter of 500 kilometers, Miranda is identical in size to Saturn's Mimas. Part of its surface is of the usual cratered type, but in addition there are deep valleys, and ice cliffs 15 kilometers high, making this the most diverse landscape encountered anywhere. Strangest of all are the enclosures that look somewhat like racetracks. The first of these showed up in the far-encounter views as a conspicuous chevron-shaped bright patch. On closer view this feature proved to consist of a patch of lighter "soil" in the midst of a roughly polygonal complex of grooved terrain. It has been named Inverness Corona, and there are two other features of the same type, Elsinore Corona and Arden Corona, which are distinctly oval.

The "racetracks," if I may use the term, appear to be areas where large pockets of subsurface rock and ice are still resettling due to gravity. Indeed, Miranda's tormented surface suggests that it is a world that has undergone more than its share of geological reshuffling, having been shattered and reformed not once but perhaps a dozen times. Each time Miranda broke up, the fragments of ice and rock tried to collect back together again, but each time they did so in a different order. In the process, the ice and rock would have been all mixed up, with the lighter ice pushing toward the outside, and the heavier rock toward the inside. As further fragments crashed together, the interior of the growing moon evidently became warm enough to melt the ice, which rose onto the

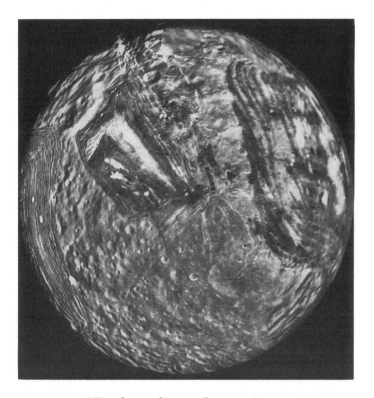

FIGURE 11.2. Miranda was discovered as recently as 1948. It displays some of the most interesting geology in the Solar System. The oval-shaped feature at the extreme left is Elsinore Corona. Inverness Corona is at upper center, and Arden Corona is at right. These jumbled areas are believed to have formed through the incomplete mixing of rock and ice fragments as the satellite broke apart and then reassembled, perhaps several times. (NASA photograph)

surface but, once there, froze so fast that the surface was left a hopeless jumble.

For all its past misfortunes, Miranda has been comparatively lucky. At least it has been able to reconstitute itself—rising up like the phoenix from its own ashes. There is firm evidence that at least one small moon lay too close to the planet to reassemble itself. When it shattered, its fragments gave rise to the planet's system of rings.

Compared to the glorious span surrounding Saturn, the rings of Uranus are a mere skeleton structure. It has even been estimated that they contain less material than is present in one of Saturn's gaps. Indeed, if all the material in the rings of Uranus were gathered together into a ball, there would hardly be enough to make a satellite 10 kilometers across. By comparison, Saturn's rings contain enough material to make several 100-kilometer satellites.

One of the most striking features about the Uranian rings is their darkness. They are, in fact, literally as black as soot, and photographing them was rather like getting a good portrait of the proverbial black cat in a coal bin at midnight. As Voyager 2 moved behind the rings, it was expected that they would become luminous by forward-scattered light, owing to the presence of fine dust, just as had been observed in the rings of Jupiter and Saturn. The Uranian rings remained dark, however, which means that they hold but little dust. The average size of the particles composing them is on the order of 1 meter across.

There are, however, sparse dust bands between the rings, which were discovered on a Voyager exposure of 96 seconds taken with the rings just 8° from the Sun and the spacecraft in Uranus's shadow to protect the cameras from glare. The dust is produced by grinding collisions between the larger ring particles, but it is rapidly cleared, for two reasons. First, the dust is relatively unresponsive to the gravitational forces of shepherd moons that keep larger chunks straitjacketed into well-behaved rings, and second, it is highly susceptible to atmospheric drag. Uranus's upper atmosphere is greatly distended, reaching all the way out to the rings, and the resulting drag efficiently sweeps dust from the rings in a few tens of days.

The extreme thinness of the Uranian rings has always been their chief claim to fame, and in their attempts to explain this, Peter Goldreich and Scott Tremaine were led to their celebrated shepherd-moon theory in 1979. Even the widest of the rings, the ε ring, is an eccentric ringlet only 60 kilometers across, so that it is no more substantial than Saturn's F Ring. Others measure on the order of a few kilometers wide.

Astronomers eagerly awaited Voyager 2's flyby of Uranus to see if it would turn up any shepherd satellites. Ten small moons were found, all within the orbit of Miranda. They have received Shakespearian names and are, in order of their distance from the planet, Cordelia, Ophelia, Bianca, Cressida, Desdemona, Juliet, Portia, Rosalind, Belinda, and Puck. Puck is the largest at 150 kilometers wide, and only two of these, Ophelia and Cordelia, are bona fide shepherds—they are 15 to 25 kilometers across and are responsible for confining the ε ring.

Presumably there must be shepherds confining the other rings, but they

FIGURE 11.3. The crescent Uranus, a dramatic
farewell shot taken by Voyager 2 from behind the
planet following its January 24, 1986, flyby. (NASA
photograph)

are bound to be very small and vulnerable to destruction through collisions with ring debris. They could not survive for long by cosmic standards, and it follows that any rings hemmed in by them would be equally
short-lived. According to current theories, the rings of Uranus cannot be
much more than 600 million years old, and the two shepherds, Ophelia
and Cordelia, may well be nothing more than the largest fragments of the
precursor satellite shattered by the impact that formed the rings.

As Voyager looked back at Uranus following its rich encounter, it captured the planet's scimitarlike crescent (fig. 11.3), an aquamarine swath

against the blackness of space. Surely this was one of the most breathtaking views of the Space Age. One can only imagine what William Herschel would have made of it.

More even than the encounters with Jupiter and Saturn, Voyager 2's encounter with Uranus was epochal, if only because so little was previously known about the planet. When Voyager 2 left Cape Canaveral back in 1977 to set out on its long journey, there were few who expected it to make it so far. Uranus seemed inconceivably remote, and Neptune, Voyager's next proposed destination, seemed beyond all possibility. As Voyager looked back at that aquamarine crescent, it was hard to believe that Uranus was now behind it, Neptune only a little more than three years ahead.

Neptune, ho!

CHAPTER TWELVE

Neptune

Before Voyager 2, Neptune was little more than a tiny bluish disk in even the most powerful telescopes, and it was only too easy to forget that in reality it was a world whose circumference was seven times that of the Earth. The most remarkable thing about it was that its existence had been worked out mathematically even before it was recognized in a telescope. This is a fascinating story, full of villains and heroes, and having many quirks and twists of fate.

The story begins in 1781 with Herschel's discovery of Uranus. As Herschel later pointed out, this discovery was hardly accidental, given the task he had set himself, which was, he said,

> never to pass by any, the smallest part of [the heavens,] without due investigation. This habit, persisted in, led to the discovery of the new planet. . . . This was by no means the result of chance, but a simple consequence of the position of the planet on that particular evening, since it occupied precisely the spot in the heavens which came in the order of the minute observations that I had previously mapped for myself. Had I not seen it when I did, I must inevitably have come upon it soon after, since my telescope was so perfect that I was able to distinguish it from a fixed star in the first minute of observation.

Yet though Herschel deserves all the credit, the prediscovery observations by Flamsteed, Mayer, and Lemonnier were not without value. Because Uranus takes 84 years to complete a circuit of the Sun, it would be a long time before it appeared again in the part of the sky where Herschel had found it. The prediscovery observations allowed the planet to be tracked back through a complete revolution, and it was fully expected that with their help an accurate orbit could be calculated.

Early orbits were worked out by Father Placidus Fixlmillner of Germany and J.B.J. Delambre of France, but oddly enough, Uranus refused to keep to its predicted course, and by 1820 the discrepancy in its orbit had become nothing less than embarrassing. That year, a fresh attack was mounted by France's Alexis Bouvard. Born in the village of Chamonix, in the valley below Mont Blanc, Bouvard spent his early years there as a shepherd, but at eighteen he went to Paris, discovered a talent for mathematics, and was soon employed in doing calculations for the great Laplace. In the days of Fixlmillner, only two "ancient" observations were available, one each by Flamsteed and Mayer. By 1820 there were seventeen. Finding that it was impossible to satisfy both the old observations and the new, Bouvard did the most logical thing and threw the old ones out. He did so in full knowledge that he was thereby imputing errors of 40, 50, and in the case of Flamsteed, even 60 seconds of arc to observers whose accuracy could generally be trusted to 10 seconds. In fairness to him, he apologized for this in the introduction to his tables: "I leave to the future the task of discovering whether the difficulty of reconciling the two systems results from the inaccuracy of the ancient observations, or whether it depends on some extraneous and unknown influence which may have acted upon the planet."

Over the next few years, Uranus continued to misbehave, and in an 1832 report to the British Association for the Advancement of Science, George Biddell Airy, then the director of the Cambridge Observatory and after 1835 the Astronomer Royal at Greenwich, noted that the discrepancy between Uranus's actual position and that calculated by Bouvard already amounted to a full 30 seconds of arc.

Airy himself apparently leaned toward the view that Newton's inverse square law of gravitation might not hold precisely at such great distances from the Sun. However, another possibility occurred to Rev. T. J. Hussey, Rector of Hayes in Kent, who on November 17, 1834, wrote to Airy about a recent conversation he had had with Bouvard. It concerned, he wrote,

> a subject I had often meditated, which will probably interest you, and your opinion will determine mine. . . . The apparently inexplicable discrepancies between the ancient and modern observations suggested to me the possibility of some disturbing body beyond Uranus, not taken into account because unknown. My first idea was to ascertain some approximate place of this supposed body em-

pirically, and then with my large reflector set to work to examine all the minute stars thereabout: but I found myself totally inadequate to the former part of the task. . . . I therefore relinquished the matter altogether; but subsequently, in conversation with Bouvard, I inquired if the above might not be the case: his answer was, that, as might have been expected, it had occurred to him. . . . Upon my speaking of obtaining the places empirically, and then sweeping closely for the bodies, he fully acquiesced in the propriety of it, intimating that the previous calculations would be more laborious than difficult; that if he had the leisure he would undertake them and transmit the results to me, as the basis of a very close and accurate sweep.

Though Bouvard lived on for another nine years, he never undertook the calculations he had discussed with Hussey. Moreover, Hussey was apparently serious when he told Airy, "your opinion will determine mine." Admittedly I am reading between the lines here, but I suspect that Hussey hoped that Airy, a first-rate mathematician, would calculate the position of the planet for him. Unfortunately, Airy had his own ideas, and his reply could hardly have been more discouraging. "It is," he wrote, "a puzzling subject, but I give it as my opinion, without hesitation, that it is not yet in such a state as to give the smallest hope of making out the nature of any external action on the planet. . . . I am sure it could not be done till the nature of the irregularity was well determined from several successive revolutions"—in other words, in several hundred years. Not surprisingly, Hussey did nothing further to pursue his idea.

Hussey was very much an amateur, but soon afterward one of the greatest astronomers of the nineteenth century, Friedrich Wilhelm Bessel of Germany, independently arrived at the same result. In 1840, Bessel assigned the problem of calculating the orbit of the putative unknown planet to a promising pupil, Friedrich Wilhelm Flemming. Unfortunately, Flemming died before he could finish, and when Bessel was himself at leisure to take up the problem, a serious illness of his own intervened.

Enter at this point twenty-two-year-old John Couch Adams, a Cornwall sharecropper's son who in 1841 was an undergraduate at St. John's College, Cambridge. While browsing in the college bookstore one day, he came across Airy's report to the British Association describing the errors in Bouvard's tables, and it fired his imagination. Before the week was out, on July 3, 1841, he jotted down the following memorandum in his notebook, which was found among his effects after his death:

Formed a design, in the beginning of this week, of investigating, as soon as possible after taking my degree, the irregularities of the motion of Uranus, which are not yet accounted for, in order to find whether they may be attributed to the action of an undiscovered planet beyond it.

At about the same time, he explained his purpose more succinctly to a fellow student: "You see, Uranus is a long way out of his course. I mean to find out why. I think I know."

Immediately following his graduation in 1843, Adams began working on the Uranus problem in earnest. Like Hussey and Bessel but independently of them, he had concluded that an outer planet was pulling Uranus from its predicted path, and for simplicity he assumed that its orbit was circular and that its mean distance from the Sun was close to that expected from Bode's law. This allowed for a rough calculation. The result was encouraging, so Adams approached James Challis, Airy's successor as director of the Cambridge Observatory, about obtaining additional Uranus observations. Challis wrote in turn to Airy at Greenwich. Airy could hardly have been more magnanimous, sending Adams all observations made at Greenwich between 1754 and 1830.

The problem of calculating the disturbance of one planet by another is notorious in celestial mechanics. It constitutes the daunting "three-body problem." Adams was trying to attack the problem, "from the rear," as the nineteenth-century historian of astronomy Agnes Clerke later wrote. His method was basically this. He began with the differences between the observed and predicted positions of Uranus, called the residuals. The residuals for certain periods of time he then set equal to the parameters describing Uranus's orbit (the mean distance from the Sun, the orientation of the orbit in space or longitude of the perihelion, the eccentricity, and so forth) in order to produce a set of "equations of condition." By manipulating these equations using approximate solutions for the unknowns, he calculated provisional elements for the planet. Finally, he could go back and see how well the observations were satisfied, given a planet that had these elements. The method amounted to one of successive approximations, and it was too much to expect an exact solution for the orbit of the unknown body. Nevertheless, Adams was confident that the location of the planet could be determined accurately enough to guide a telescope to its discovery—what Hussey had only dreamed of in 1834.

By September 1845, Adams had a solution—a set of orbital elements for the unknown planet and in addition its heliocentric longitude, or the approximate point in the sky where he believed it would be found. He at once turned the results over to Challis, who was in charge of the 12-inch Northumberland refractor at Cambridge Observatory. Rather than mounting an immediate search, as might be expected, Challis pleaded that he had "more important" work to do and suggested that Adams contact Airy with his results. He later confessed his reasons for this. It seemed, he wrote, "so novel a thing to undertake observations in reliance upon merely theoretical deductions, and that while much labour was certain, success appeared very doubtful." This attitude is perhaps understandable, given that Challis was always chronically overworked. As we now know, however, the planet lay at a point in the sky only 2° from where Adams put it.

At this time Adams was planning a return to his home at Lidscot, Cornwall, and offered to deliver the calculations to Airy personally. Challis furnished him with a letter of introduction and Adams set off. Unfortunately, on reaching Greenwich, Adams found that Airy was away in Paris. After his return, Airy wrote Challis suggesting that Adams drop by again the following month. Adams obliged, in fact calling upon him twice. The first time Airy was out, but Adams promised to return in the afternoon. Unfortunately, his intention was never communicated to the Astronomer Royal, and when he returned he was given the brush-off by Airy's butler, who told him that the Astronomer Royal was at dinner and could not be disturbed. (Airy had the unusual habit of dining punctually at three in the afternoon.) Adams could only sketch out his results, and, leaving these behind, he departed Greenwich feeling snubbed.

Airy was not uninterested in Adams's results, but one question nagged at him. This was the so-called problem of the radius vector. As early as the 1830s, Airy himself had found that Bouvard's tables were off not only in predicting Uranus's longitude, or position in the sky, but also in its prediction of the planet's distance from the Sun. This distance, known as the radius vector, always came out too small. Airy wrote to Adams to inquire whether his theory of an outer planet could account for this error as well. Here Adams made a serious error. Whether because, as he later claimed, he saw the question as trivial or, as Challis thought, because of procrastination, he never bothered to respond. This effectively brought the correspondence to an end, for Airy concluded, as he later explained, that Adams's silence made it "clearly impossible for me to write to him again."

Airy had an obsessive-compulsive personality. Though he deserves credit for bringing order into the affairs of the Royal Observatory, he was often known to overdo it—stories are told of his making the rounds even on cloudy nights to make sure observers were at their posts, and he once spent a whole afternoon in the observatory's cellar labeling a stack of empty boxes "Empty." The radius vector problem was clearly a detail that had to be tidied up, but in fact it was rather more than this. Airy saw the problem as nothing less than the *experimentum crucis*—on it, he believed, turned the continued acceptance or the rejection of the law of gravitation. As we have seen, Airy had to be convinced that the inverse square law still held exactly at such great distances from the Sun, and in Airy's defense, even Adams later had to admit that his conclusion about the radius vector had been somewhat "hastily inferred."

But Airy had yet another reason for not taking immediate action on Adams's results. "I think," he wrote, "that the results of algebraic and numerical computations, so long and so complicated as those of an inverse problem of perturbations, are liable to many risks of error in the details of the process." There is no doubt some truth in this, and Agnes Clerke's point is also well taken: "The difficulty of determining the perturbations produced by a given planet is small compared with the difficulty of finding a planet by its resulting perturbations. Laplace might have quailed before it; yet it was now grappled with as a first essay in celestial mechanics." This was Airy's assessment of the situation too. Frankly, he lacked confidence in Adams's abilities, which were unproved.

If the inverse problem of perturbations was one that was "liable to many risks of error in the details of the process," one man's results could be distrusted. What, though, was the probability that two men, working entirely independently through the same long chain of calculations, would arrive at nearly the same result? Yet that is just what happened. Knowing nothing of Adams's labors, a French mathematician, Urbain Jean Joseph Leverrier, tackled the problem along very much the same lines and arrived at a position differing from Adams's by less than a degree.

Ten years Adams's senior, Leverrier had begun his scientific career as a protégé of the chemist J. L. Gay-Lussac. He had gone over to astronomy in 1837 when Gay-Lussac, having to choose between Leverrier and another promising young chemist for a single vacant position in the École Polytechnique, neatly solved the dilemma by recommending Leverrier for a position of equal rank that had opened up in astronomy. This seems to have been satisfactory to all parties concerned, for Leverrier wasted no time in getting started on the outstanding problems in celestial mechanics. In the summer of 1845 his attention was directed to the problem of Uranus by François Arago, the director of the Paris Observatory, and he gave it his highest priority. (By that time, of course, though unbeknownst to anyone outside England, Adams already had the solution within his grasp.)

Reviewing Bouvard's tables, Leverrier concluded that they were, as Morton Grosser puts it, "a shoddy patchwork of errors." He corrected the errors, but there still remained a large discrepancy between Uranus's observed and predicted paths. This led Leverrier to the same conclusion Adams had reached: an outside planet was drawing Uranus off course. On June 1, 1846, he presented his complete findings to the French Academy of Sciences, including his prediction of the planet's position in the sky.

When Airy received Leverrier's memoir three weeks later, he was stunned. As might be expected, he immediately wrote to Leverrier about the problem of the radius vector. Here Leverrier's ebullience contrasted with Adams's sulking. "The radius vector is corrected," he announced, "without having to be dealt with independently. Excuse me, Sir, for insisting on the point." At the same time Leverrier suggested that Airy mount a search for the planet and offered to send his latest position as soon as he had calculated it. Airy never replied. He later excused his inaction by saying that he was just about to depart for the Continent on other business. His imminent departure did not, however, prevent him from writing to Challis on July 9, encouraging him to begin searching with the Northumberland refractor: "You know that I attach importance to the examination of that part of the heavens in which there is a possible shadow of reason for suspecting the existence of a planet exterior to Uranus." Perhaps reflecting on the impression that the phrase "a possible shadow of reason" was likely to have on the dilatory Challis, Airy wrote again, more peremptorily, four days later: "In my opinion, the importance of this inquiry exceeds that of any current work, which is of such a nature as not to be totally lost by delay."

Airy suggested that Challis sweep a generous band of the zodiac 30° long by 10° wide and centered on the planet's position as predicted by Leverrier. In order to be especially thorough, Challis decided that, instead of searching stars only to the ninth magnitude, which was Adams's expectation of the brightness of the planet, he would include stars down to the eleventh magnitude. This meant wading through six times as many stars. He got underway on July 29, recording the position of each star in the zone, and on August 12 he decided to compare his observations of that night with those of July 30, just to be certain of his methods. He stopped his cross-check at star number 39. Had he continued for ten more stars, he would have captured the planet.

Challis continued ploddingly in this course, reporting to Airy in early

September: "I get over the ground very slowly . . . and find, that to scrutinise, thoroughly, in this way the proposed portion of the heavens, will require many more observations than I can take this year." It is hard to imagine that he felt the least enthusiasm for this work; he was only doing his duty. It is generally supposed that he was the only searcher, but not so. In London, John Russell Hind, who would later become a well-known discoverer of asteroids was sweeping for Leverrier's planet with a 7-inch refractor (despite corresponding with Airy, he knew nothing of Adams's earlier work). In addition, a brief attempt was apparently made at the Paris Observatory.

Others, too, had opportunities that they did not seize. As Challis was getting underway at Cambridge, Sir John Herschel suggested to his friend Rev. William Rutter Dawes that he ought to make a search for the planet. Instead of acting on the suggestion, however, Dawes passed it along to William Lassell, whose 24-inch reflector was the most powerful telescope in England at the time. Unfortunately, Lassell was laid up with a sprained ankle, so he too missed his chance.

Then there is the whole question of why Adams and Leverrier never bothered to look for the planet themselves. The lapse can be blamed, I suppose, on specialization. They were theoreticians, not observers, and Camille Flammarion, who later worked under Leverrier as an assistant at the Paris Observatory, went so far as to write after the planet's discovery: "The author of the calculation himself, the transcendent mathematician, did not give himself the trouble to take a telescope, and look at the sky to see whether a planet was really there! I even believe that he never saw it." Similarly, though Adams had completed his calculations by September 1845, he left the matter wholly in others' hands, so for ten months no one searched for the planet despite the fact that, as we now know and as Adams himself had implied, it would have been within reach of a simple 2-inch telescope.

Eventually, no doubt, Challis's plodding search would have turned up the planet, but now the pace of events began to quicken. On September 10, 1846, Sir John Herschel, sensing that the discovery of a new planet was indeed imminent, addressed the British Association for the Advancement of Science with stirring words: "We see it as Columbus saw America from the shores of Spain. Its movements have been felt, trembling along the far-reaching line of our analysis with a certainty hardly inferior to ocular demonstration." At that time Herschel fully anticipated that the priority for the discovery—both in terms of the calculated position and the ocular demonstration itself—would redound to the glory of Cambridge. Adams, Airy, Challis, and Herschel himself were all Cambridge men.

But the initiative passed to others. On September 18, 1846, Leverrier, knowing nothing of the efforts afoot in England, wrote to Johann Gottfried Galle, an assistant astronomer at the Berlin Observatory, to thank him for a copy of his dissertation, which had gone a year unacknowledged. This was, however, only a lead-in to his true motive: "to find a persistent observer, who would be willing to devote some time to an examination of a part of the sky in which there may be a planet to discover." Galle proved to be as enthusiastic as Leverrier could have wished, and he received permission from the observatory's director, J. F. Encke, to mount the search.

Assisted by an eager young graduate student in astronomy, Heinrich Louis d'Arrest, Galle went to the dome of the observatory's 9 1/2-inch Fraunhofer refractor on the evening of September 23, 1846. He pointed the fine instrument toward the point in the sky that Leverrier had indicated: right ascension 21 hours, 46 minutes, declination −13° 24′. At first he tried to make out a tiny disk among the stars, but without success. At this point, d'Arrest suggested using a star chart. The most accurate chart of this region of the sky had been compiled by Carl Bremiker for the Berlin Academy and printed a year earlier, but it had not yet been made available to observatories elsewhere, for no better reason than that the next map in the series was nearly finished and the academy had decided to mail both together in order to save postage. Had Challis had this map in July, he would undoubtedly have claimed the planet then—indeed, an earlier map in the same series *was* in his possession, and it covered a small part of his search zone, including, it turns out, the area where the planet was then hiding. Challis, however, never used it for this purpose.

D'Arrest pored over the Bremiker star chart while Galle again took his place at the eyepiece. One by one Galle called out the positions and brightnesses of the stars in the field of view while his assistant checked them off against the chart. After about an hour, Galle called out the position of an eighth-magnitude star. D'Arrest hesitated, then shouted: "That star is not on the chart!"

Close inspection revealed a tiny disk, which Galle measured the next night. His value, 2.7 seconds of arc, is very close to the currently accepted value of 2.5. Galle wrote immediately to Leverrier: "The planet whose position you have pointed out *actually exists*." It lay only 55 minutes of arc—less than twice the diameter of the full moon—from Leverrier's position and 1.5° from that indicated by Adams.

Challis's cup was bitter indeed. Not only had he missed the planet by failing to compare his observations in July, but on September 29—the night before the news from Berlin reached England—he swept across a star that seemed to have a disk. It was, of course, the planet, but instead of examining it at once with a higher power, he characteristically put the chore off until the next night, and then he found another excuse: the Moon was in the way. The Moon's proximity did not, however, prevent Hind, who had just received the news from Berlin, from making out the planet with a 7-inch refractor in London. Thus Hind, not Challis, became the first Englishman to recognize the planet, and Challis lost his final chance for redemption.

With good reason, Leverrier felt that he was entitled to name the planet, and he at first proposed Neptune. Almost at once he changed his mind, however, and schemed to have the planet named for himself. In the published edition of his *Recherchés sur les Mouvements d'Uranus,* he apologized for calling Uranus by that name, vowing in future publications to consider it "my strict duty to eliminate the name Uranus completely, and to call the planet only by the name *Herschel.*" Clearly this was, as William Graves Hoyt has noted, a "not-so-subtle ploy to gain acceptance for his proposal" to name the new planet Leverrier. Of course, the name was not adopted.

Up to this time, Adams's independent—and in fact earlier—investigation remained unknown outside of the small circle in England. The first public announcement of it came from Sir John Herschel in a letter written to the London *Athenaeum.* In it he wrote:

> The remarkable calculations of M. Le Verrier, if uncorroborated by repetition of the numerical calculations by another hand, or by an independent investigation from another quarter, would hardly justify so strong an assurance as that conveyed by my expression [to the British Association]. . . . But it was known to me, at the time (I will take the liberty to cite the Astronomer-Royal as my authority) that a similar investigation had been independently entered into . . . by a young Cambridge mathematician, Mr. Adams;—who will, I hope, pardon this mention of his name.

Herschel's letter unleashed a storm of controversy. A shocked Leverrier probed Airy: "Why has Mr. Adams kept silent? . . . Why did he wait until the planet was seen in the telescope?" The full answer, when it became known, was very embarrassing to both Airy and Challis. Patriotic feelings were roused on both sides of the Channel. The French cried foul, and in England Airy and Challis were violently attacked.

The hard feelings were long in dying out—as late as the 1890s, when it was proposed that Airy should be commemorated in Westminster Abbey, the idea was voted down because of lingering memories of the Neptune affair. But at least it is pleasant to record that when Adams and Leverrier actually met, at Sir John Herschel's estate at Collingwood in June 1847, it was on terms of the utmost respect and "perfectly free from jealousy." Leverrier has a reputation for being an extremely ill-tempered man, but this incident can only make one wonder whether such a reputation is justified.

Soon after the announcement of Neptune's discovery, telescopes the world over were directed toward the newest member of the Solar System. Among them was William Lassell's 24-inch reflector, erected earlier that year at Starfield, his residence near Liverpool. As a young man, Lassell had entered on a career as a brewer, and though he denied having "much taste or inclination for trade," he succeeded well enough to become a man of means so that he was able to devote his later years to his real interest—astronomy. He had, as we have seen, missed a chance to search for the planet himself. He learned of its discovery in a letter from Sir John Herschel dated October 1, 1846, in which Herschel added almost as an afterthought: "Look out for satellites with all possible expedition!!"

Lassell's large reflector was the first such instrument to be furnished with an equatorial mount, and he was justly proud of it. He turned it toward the planet "Leverrier" on October 2 and on the very next night first recorded his suspicion of a ring around the planet. He made a sketch on October 10 and also observed a "star" whose "close situation . . . and minuteness," as he noted at the time, "occasioned my strong suspicion that it may be a satellite."

Because of poor weather in England, the satellite, Triton, was long in being confirmed. Lassell did not finally succeed until July 1847. On the whole he seems to have been rather more confident of the ring, and in late 1846 he published a first account of his observations. Following up on Lassell's cue, Hind, who had hitherto never once suspected that the planet's disk was other than round, found that "Leverrier presents an oblong appearance." Challis, using the 12-inch refractor at Cambridge, also reported a ring.

For a few months the ring seemed to be established beyond doubt, but other skilled observers could find no trace of it, and in time it came to be forgotten. The closest Lassell himself came to a retraction was at the end of 1852, when he noted that, whatever its cause, the ring was perhaps more "intimately related to the telescope" than to Neptune.

We now know that Neptune does have several faint rings, but as with the sightings by William Herschel of a ring around Uranus, there is no possibility that Lassell saw anything of the genuine article. The pressure to find something with his powerful telescope in the wake of the exciting discovery, combined with indifferent seeing conditions in England that fall, may well account in part for Lassell's observations. The main culprit, however, seems to have been flexure in the giant speculum-metal mirror, which seems to have played a role in producing optically elongated images, which Lassell interpreted as a ring tilted nearly edge-on. As for the "confirmations" of Hind and Challis, these only prove the power of suggestion, which often leads observers to see what they expect to see.

The name Neptune, for the sea-god of classical mythology, was adopted by the astronomers of Europe—Leverrier's first choice being preferred to that which his vanity had later suggested. The name is fitting, given the planet's sea-blue tint in the telescope, which even a modest aperture will show to be different from greenish Uranus.

As difficult as Uranus was as an object of study, Neptune seemed even more out of reach. Bands and spots were glimpsed by visual observers from time to time, but it was difficult to trust these sightings. After all, such appearances were only to be expected by analogy to the sunward giants, and it is a well-known fact that if one looks hard enough, one is apt to see something eventually—even if it isn't there. I have had the chance to examine the planet with various telescopes, including some rather large ones, but have never made anything out. True, I do not claim to have exceptional eyesight or ability as an observer, but even Stephen J. O'Meara, who has a well-earned reputation for eyepiece clairvoyance, found Neptune in August 1988 to be only "an ice-blue world with hazes no more intense than a breath on a cold windowpane" even though he was observing with a very large telescope, the 60-inch reflector at Mt. Wilson.

Because of its remoteness, Neptune long kept its secrets, and before Voyager 2 there was little accurate information. In size it was nearly an exact twin of Uranus, with a diameter of 49,528 kilometers. From changing cloud patterns observed in the infrared, its rotation period was set at 17 to 18 hours, also similar to that of Uranus, but its axis of rotation was orthodox, being tipped only 29° from the perpendicular. Unlike Uranus and like Jupiter and Saturn, Neptune was known to give off more energy than it receives from the Sun. This was thought to explain the greater activity of its atmosphere compared with that of its bland twin, Uranus.

Stellar occultation measurements from the Earth during the early and mid 1980s suggested that Neptune was surrounded by a set of broken rings, or ring arcs (see below). Also, two satellites were known, Lassell's large moon Triton and tiny Nereid, discovered photographically by G. P. Kuiper in 1949. But that was all. One could only ask with the nineteenth-century astronomer Rev. T. W. Webb: "Who can say how grand a spectacle this inconspicuous globe might present on nearer approach?"

The nearer approach finally came on August 25, 1989, when Voyager 2 swooped to within 4,905 kilometers of the planet's upper cloud deck at a point not far from the north pole. The spacecraft was then swung toward Triton by Neptune's gravitational pull, passing within 40,000 kilometers of it six hours later.

Light intensity levels at Neptune were only 40 percent of those at Uranus, or 1/900th of those at the Earth. This meant that photography was bound to be difficult, with exposure times of 15 seconds or more needed. As at Uranus, the spacecraft's motion relative to its target had to be compensated for by carefully panning the spacecraft—what Voyager scientists dubbed their "anti-smear campaign."

In far-encounter photographs, Neptune already showed a number of striking cloud features on its sea-blue disk. The first to be disclosed was a brilliant white cloud in the southern hemisphere, so conspicuous that it was visible in the best Earth-based images. Nearer views showed that it consisted of cirrus clouds of methane ice skirting along the southern edge of a tremendous storm that became known as the Great Dark Spot (fig. 12.1). This was the outstanding feature of the whole disk, with a width equal to that of the Earth. Bearing a striking resemblance to the Great Red Spot of Jupiter, it is an anticyclonic system, churning around itself once every sixteen hours. In August 1989 it was located at latitude 20°s, but it drifts wildly in longitude and latitude.

The cirrus clouds fleeting along its southern edge form as updrafts lift methane high into the atmosphere, where it freezes out. Thus the clouds are somewhat analogous to the orographic clouds of the Earth or Mars, which form as moist air is uplifted over mountains. On Neptune the effect is produced by pressure and temperature variations in the giant storm system.

The methane cirrus clouds form at heights of some 50 kilometers above the methane cloud deck. This is known directly, because the Voyager

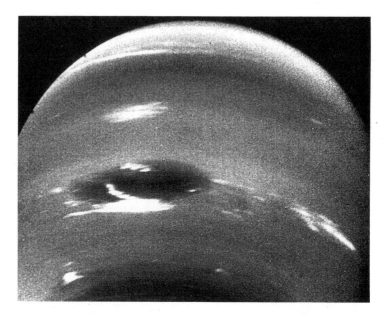

FIGURE 12.1. This Voyager 2 image of Neptune was obtained on
August 21, 1989, from a distance of 6.1 million kilometers. The
dominant feature is the Great Dark Spot, with its attendant wreath
of bright methane cirrus clouds. Also shown, at lower left, is the
bright feature known as Scooter. (NASA photograph)

cameras captured some of these clouds in the northern hemisphere near
the day-night terminator, where their thin silvery wisps cast measurable
shadows on the sea-blue ocean below. Incidentally, this was the first time
that relief was directly observed on one of the giant planets. The colorful
cloud decks of Jupiter and Saturn had appeared completely flat.

A smaller dark spot, known as D2, was found at latitude 55°s. Then
there was the brilliant white "Scooter," at latitude 42°s. Both are shown in
figure 12.1. Scooter appeared to be an extension of an ammonia or
hydrogen sulfide cloud deck that lies some 30 to 50 kilometers below the
blue methane clouds.

The rotation period of the Great Dark Spot varied between 18.28 and
18.38 hours, "Scooter" rotated in 16.75 hours, and so on. The behavior of
D2 was of special interest. When first seen, it was at latitude 55°, and its
period was 16.0 hours, but it underwent several changes of speed with
corresponding shifts in latitude. As its period slowed to 16.3 hours, it

moved northward to latitude 51° and then sped up again, returning to its previous latitude but now with a shortened period of 15.8 hours.

The rotation period of the planet's core, 16.11 hours, was found from radio observations of bursts from the interior. Relative to the core, then, the large-scale atmospheric features in the equatorial region are sweeping westward at 1,200 km/hr, and smaller features at twice this speed. This makes Neptune the windiest planet in the Solar System.

The vigor of its winds was rather startling, considering that Neptune is so remote from the Sun. The periods of rotation are longer at the equator than near the poles and consistently lag behind the rotation of the deep interior. According to a recent theory, Neptune's interior rotates as a solid body, and at 2,000 to 3,000 kilometers below the visible cloud tops a transfer of angular momentum takes place between the solid body and the outer gaseous envelope. Because angular momentum is conserved, as a parcel of gas rises it falls behind the rotation of the solid interior, and this results in the strong easterly jet observed in the equatorial region there. The important point is that the winds of Neptune are not chiefly driven by the Sun.

The most striking, and unexpected, resemblance between Neptune and Uranus is that they are alike magnetically. Uranus's magnetic field, as we have seen, is tilted by some 60° to its axis of rotation, a fact that used to be explained by invoking a tremendous off-center blow from another planet-sized body late in the formation of the Solar System. There seemed every reason to expect that on Neptune things would be normal. Not so. Despite Neptune's orthodox axis of rotation, Voyager found that the magnetic field is tilted to this axis by 47°. Consequently, as at Uranus, the magnetic field wobbles wildly as Neptune spins. The cause of this state of affairs is now thought to be that both planets are "oblique rotators," generating their magnetic fields through convection of electrically conducting material within a thin shell near the surface rather than in a molten core, as on the Earth.

Voyager confirmed that Neptune has a set of rings, and in the process also shed light on the puzzling "ring arc" observations that had been made from the Earth. The outermost ring that Voyager found lies at a distance of 62,900 kilometers from Neptune's center, and it does indeed contain three bright ring arcs, all concentrated in the same segment of the ring, with wispy material between forming a continuous circle around the planet. Recent calculations have shown that the ring arcs are clumps of material produced through resonance interactions with the inner moon Galatea.

Another narrow ring lies at a distance of 53,200 kilometers from the planet, and two more diffuse rings and a sheet of dust extend from 59,000 kilometers all the way to the planet's cloud tops.

As with Saturn and Uranus, the shepherd-moon theory has been invoked to explain the existence of narrow rings, and near the inner edges of the two main rings Voyager did indeed find shepherd moons, Despoina and Galatea, which keep the particles from straying inward. However, no shepherds were found at the rings' outer edges. Shepherd moons may exist there, but if so, they have diameters smaller than the 12-kilometer limit identifiable by Voyager's cameras.

Neptune's ring system contains a dearth of millimeter- to meter-sized particles, but there is plenty of dust. Thus the rings became luminous as Voyager looked back at them from behind the planet. In all, the rings contain only enough material to make up a moonlet about 5 kilometers across, and compared to the spectacular system of Saturn or even Uranus, they are ancient and weathered.

In addition, Voyager found half a dozen small moons. All but one lie close enough to the planet so that, if broken apart by a random comet strike, they would form new sets of rings. Apart from the two shepherds mentioned above, there are two moons even closer to the planet, Naiad and Thalassa. Another, Larissa, orbits 10,000 kilometers beyond the outermost ring. All have diameters in the 50- to 200-kilometer range, so their destruction would lead to a system rivaling Saturn's. The one new moon too far out to form rings is the largest of the set. It has received the name Proteus and is 400 kilometers across. It orbits the planet at a distance of 117,600 kilometers from its center.

Incidentally, the half dozen new Neptunian moons brought the total number of Voyager moons to twenty-five. Putting this into perspective, of the Solar System's sixty-one known moons, forty percent were discovered by the Voyager spacecraft.

Neptune and its rings were marvelous, but to my mind the highlight of the encounter was Triton. I had often seen it in my 12 1/2-inch reflector as a mere speck of light rather easier to make out than Uranus's Oberon or Titania. But no one could ever have guessed what a world of wonders awaited us there (see fig. 12.2).

At one time it was thought that Triton might well be the largest satellite in the Solar System, but its diameter has now been reliably measured at just 2,700 kilometers—rather less than Earth's Moon. Its orbit is markedly abnormal. Though nearly circular, with a radius of 354,800 kilome-

FIGURE 12.2. A Voyager 2 photomosaic of Triton. The south polar cap is the bright, mottled area at left. The elongated dark patches consist of sooty deposits produced from still-active volcanic processes. The so-called cantaloupe terrain can be seen to the right (equatorward) of the south polar cap margin. (NASA photograph)

ters, it is tilted by an angle of 23° to Neptune's equatorial plane. Moreover, Triton travels in a retrograde or "backward" sense around its orbit. This behavior is not uncommon among the most minute satellites of the Solar System, most of which were undoubtedly captured asteroids, but it is unique among the large ones.

The only possible explanation seems to be that Triton, too, is a captured object. This would also account for its unusual composition. Unlike the icy satellites observed by Voyager at such great distances from the Sun, Triton contains as much rock as Io and Europa. Thus it must have formed in a different part of the Solar System and only later been snared gravitationally by Neptune.

Around Triton's south pole is a large, reddish polar cap, believed to consist mostly of nitrogen and methane. In the complex 600-year cycle of Triton's seasons, sunlight is now shining more directly on the south pole than it has for centuries, causing the cap to retreat at various spots along its border. Even so, Triton still reflects so much sunlight back into space that its temperature is the lowest of any body yet visited by spacecraft—only 38°C above absolute zero (−273°C).

Extending toward the equator from the south polar cap lies a richly varied terrain. On Triton's trailing edge the texture is remarkably like that of a cantaloupe rind. This may be a region where gases and fluids bubbled to the surface long ago. It is the youngest region on Triton, with few impact craters. As one moves toward Triton's leading edge, the cantaloupe-rind terrain gives way to frozen lakes surrounded by multiple terraces. These were formed, it is thought, during past eras of flooding by liquid water.

Clearly, Triton has been geologically active during much of its history. When first captured by Neptune, it must have followed a highly eccentric orbit, but gradually this shape was tamed into a near circle by tidal forces, and in the process the moon's interior melted, producing volcanism that completely remodeled the surface.

Even now the moon is far from dead. Mottling the south polar cap are numerous dark streaks, which suggest that material has recently erupted from below. Moreover, Voyager observed two active geyser plumes. They consist of material ejected from the surface to a height of 8 kilometers and then blown downwind for distances of up to 150 kilometers in Triton's nitrogen and methane atmosphere, whose surface pressure is only 1/70,000th of that of the Earth.

What fuels the powerful eruptions? Almost all the nitrogen gas on Triton is frozen onto the surface in the form of an almost transparent glasslike jacket covering the moon to a depth of 1 to 1.5 meters. Dust particles consisting of dark organic material are embedded in this jacket of frozen atmosphere. When directly exposed to midsummer sunlight, these dust particles heat the underlying surface through the greenhouse effect. As soon as the temperature rises above the boiling point of nitrogen, nitrogen gas and dark organic material are released at explosive speeds of perhaps 100 meters/sec.

Both of the geyser plumes observed by Voyager lie near the current subsolar point, and there can be little doubt that geysers are erupting constantly on Triton—a surprising vitality, considering that this was a world that, before Voyager, was fully expected to be in the suspended animation of deep-freeze.

Finally, something should be said of Nereid, the tenth moon of Neptune and aside from Triton the only one discovered from the Earth. It measures 340 kilometers across, but its most remarkable feature is its orbit, which is among the most eccentric and sharply inclined in the Solar System. Incredibly, Nereid's distance from Neptune ranges all the way from 1,390,000 kilometers to 9,640,000. Needless to say, there is every reason to believe that it is a captured body. Unfortunately, Voyager's closest approach to it was 4,700,000 kilometers, so it could not be studied in detail.

Once Neptune was only the dream of Adams and Leverrier, an unknown mass caught in the web of their mathematical analysis. Then Galle laid eyes on it, confirming its existence, and for a century and a half it was as he had seen it: a dim and impossibly remote blue globe in the circle of the eyepiece, companioned by the glint that Lassell first captured in his large reflector in the weeks following the planet's discovery.

Now Voyager has brought the planet home to us, and it haunts our imaginations as a floating ocean of blue flecked here and there with white clouds and surrounded by wispy rings and strange moons. Not even one Neptunian year has passed since Galle first saw it. Who can say what science will have brought to pass by the time Neptune makes another round?

CHAPTER THIRTEEN

Pluto

No one could deny the spectacular success of Adams and Leverrier in predicting the position of Neptune—Adams was off by 1.5° and Leverrier by only 55 minutes. But in other respects their postulated orbits were rather wide of the mark. In particular, Neptune's actual distance from the Sun, 30.25 AU, which was established soon after its discovery in 1846, was much less than predicted by either Adams or Leverrier (36.15 and 37.25 AU, respectively). Because of this discrepancy, Harvard mathematician Benjamin Peirce went so far as to declare in 1847: "The planet Neptune is not the planet to which geometrical analysis had directed the telescope; its orbit is not contained within the limits of space which have been explored by geometers searching for the source of the disturbance of Uranus; and its discovery by Galle must be regarded as a happy accident."

Peirce's argument turned on the fact that a planet located at a distance of 35.3 AU would have a period of revolution 5/2 as long as that of Uranus. But the existence of such a commensurable period, Peirce maintained, would introduce such marked irregularities into the motions of both bodies as to produce nothing less than "a break in the continuous law" by which Adams and Leverrier had made their inferences. Another interesting feature Peirce pointed out was that Neptune's actual location was near another commensurable period with Uranus: the ratio of the periods of revolution—164.8 years for Neptune to 84 years for Uranus—was nearly 2:1.

Despite Peirce's arguments, other astronomers found the idea that Neptune's discovery had been nothing more than a "happy accident" unpalatable to say the least. Leverrier himself attempted to refute it. So did Sir John Herschel, who argued that because Uranus and Neptune had passed conjunction in 1822, the perturbations between them had been near maximum at about that time, so Adams and Leverrier's prediction of the direction of the disturber had been precise despite the other inaccuracies in the postulated orbits.

On one point, at least, there was agreement. There seemed no reason to believe that Neptune marked the outer terminus of the Solar System— indeed, Leverrier had said as much to Galle as early as October 1, 1846. But though it was found that the French astronomer Lalande had twice recorded Neptune in May 1795 without recognizing that he had a planet in his sights, the arc Neptune had traversed was still too short to be useful in identifying the position of a possible outlier. This meant that astronomers once again had to rely exclusively on the motion of Uranus in pointing them in the direction of planets still unknown.

By including Neptune in the calculations, the residuals of the motion of Uranus were, of course, dramatically reduced, yet as early as 1877 David Peck Todd of Amherst College in Massachusetts found a small remainder. Using a method in which the peaks of the graph of these residuals were taken to indicate conjunctions of an outer planet with Uranus, Todd concluded that there was a planet orbiting the Sun at a distance of 52 AU and in a period of 375 years. He even worked out a longitude for it, placing it in the constellation Virgo. On thirty nights between November 1877 and March 1878 he searched for the planet with the 26-inch refractor at the U.S. Naval Observatory, but in vain.

A year later the French astronomer Camille Flammarion observed that the aphelia of periodic comets and meteor swarms tend to cluster around certain distances from the Sun. Thus he found thirty-three comets with aphelia near the orbit of Jupiter, while Neptune had a family of six. He added that two comets had aphelia at a distance of 47.5 AU and suggested that this might indicate the existence of a planet at that distance.

Several other astronomers made planetary predictions, but by far the most illustrious were two Americans, Percival Lowell and William Henry Pickering. Lowell, of course, is best remembered for his observations of the "canals" of Mars and his deductions therefrom about intelligent life on that planet. But during his Harvard days he had acquired a taste for celestial mechanics from Benjamin Peirce, whom he acknowledged as his mathematical "master." The respect was mutual, for Peirce in turn had called Lowell "the most brilliant mathematician of all those who have come under my observation." Lowell had acquired from Peirce a firsthand acquaintance with the facts of the Neptune controversy, and he was also aware of Flammarion's idea of deducing the existence of outer planets from comet aphelia, writing as early as 1902 that "not only are comets a part of our system now recognized, they act as finger-posts to planets not yet known."

Lowell expressed these opinions in a somewhat casual manner, but by 1905 he had begun to take a serious interest in the subject. In the midst of his other work, especially on Mars, Lowell started a photographic survey of the "invariable plane"—the mean orbital plane of the known planets, which was the most likely place to find an unknown planet orbiting the Sun. He delegated the photographic work to a series of assistants—John C. Duncan, Earl C. Slipher, and Kenneth P. Williams—who exposed several hundred plates over the next three years with a 5-inch photographic refractor. They photographed a given region of the sky twice at an interval of two weeks and then shipped the plates to Boston, where Lowell lived when he was not actually observing. At first, Lowell's method of inspecting them was simply to lay the plates of a pair atop one another and to examine them with a hand magnifier.

Eventually Lowell became convinced that without a calculated position to assist in directing the telescope to approximately the right part of the sky, there was little chance of success. Already in 1905 he had hired an assistant to compute the residuals of the outer planets as the first step toward obtaining such a position. Thoroughness initially counted more with Lowell than speed. But in November 1908 he learned that Pickering was also in the race—indeed, that he was well ahead of him. From a graphical analysis of the residuals of Uranus like that used by Todd, Pickering announced to the American Academy of Arts and Sciences in Boston his prediction of "Planet O" at a distance from the Sun of 51.9 AU and with an orbital period of 373.5 years. Brief photographic searches of the position indicated by Pickering were mounted, but the planet failed to show itself.

Pickering's appearance on the scene gave added urgency to Lowell's quest. He at once tried his own hand at the graphical method of prediction. The results were never published, however, and in any case, by the summer of 1910 Lowell had decided to abandon such crude methods—which even Pickering admitted partook not a little of the "rule of thumb"—in favor of the rigorous analytical method Leverrier had used in the search for Neptune.

Indeed, from 1910 until his death Lowell became increasingly absorbed in the quest for the unknown planet, which he called Planet X. Harassed by critics of his Martian theories, Lowell hungered desperately for some sort of vindication. The search for Planet X became the great obsession of his later years, as the Martian "canals" had been of his earlier. It was the mad chase of a man dominated by an idée fixe. The mathematical problem he faced was, however, infinitely more difficult than that Adams and Leverrier had solved, insofar as the outstanding error in the residuals of Uranus of 133.0 seconds of arc had, with the inclusion of Neptune's influence, shrunk to a mere 4.5 seconds. Lowell did his utmost to force this unpromisingly small figure to yield the position of a planet. At times he worked himself to utter exhaustion. The technical details of his mathematical investigation need not concern us here. Suffice it to say that each time Lowell revised his position in Boston, he immediately wired his assistants in Flagstaff, telling them to expose a new set of plates of that region of the sky, and his positions sometimes changed from week to week. His first calculations put the planet in Libra, but over the next several years the favored position moved to eastern Taurus, in the heart of the Milky Way's myriad of stars. We now know that two plates taken in 1915 by Lowell's assistant Thomas Gill with a 9-inch photographic refractor borrowed from the Sproul Observatory actually showed the planet, but it went unnoticed, lost in the thicket of stars.

No doubt with an eye to his critics, Lowell kept his search as secret as possible, but finally in January 1915 he was ready to make public the results of his mathematical investigation in an address to the American Academy of Arts and Sciences. Alas, the address created hardly a stir. The Academy even declined to publish the paper. When it finally appeared nine months later, it was published at Lowell's own expense as his *Memoir on a Trans-Neptunian Planet*.

Among the results Lowell presented was a refutation of Peirce's arguments against the validity of Adams and Leverrier's prediction. Lowell found that "there is no discontinuity . . . at the points of commensurable period, the function passing through them without a break." In discussing his own calculations, he closed with words that somewhat uncharacteristically convey a note of humility about his methods:

> Owing to the inexactitude of our data . . . we cannot regard our results with the complacency of completeness we should like. . . . The fine definiteness of positioning of an unknown by the bold analysis of Leverrier or Adams appears in the light of subsequent research to be only possible under certain circumstances. Analytics thought to promise the precision of a rifle and finds it must rely on

the promiscuity of a shot gun after all, although the fault lies not more in the weapon than in the uncertain bases on which it rests.

In summary, Lowell's final calculations assumed a mass for his planet seven times that of the Earth, and he put its position in eastern Taurus near the border with Gemini.

After the publication of his *Memoir,* Planet X apparently faded from the foreground of Lowell's thought. The search continued for a while yet—the last entry in the logbook of the 9-inch photographic refractor, made in July 1916, reads "Lunch," which suggests a project only temporarily broken off. But it is only too clear that by then Lowell himself had given up, and following his death from a massive stroke in November 1916, thirteen years would pass before another plate would be exposed at his observatory in the search for Planet X.

There was a brief interlude in 1919 when Pickering announced a revision of his 1908 prediction for Planet O. Using a 10-inch refractor at Mt. Wilson, Milton Humason exposed several plates of the region suspected of holding the postulated planet, but once again the results were negative. One suspects that, in any case, Pickering's credibility was becoming rather strained by then, as he had for sometime been arguing for the existence of not one but several trans-Neptunian planets, whose orbits he tended to revise drastically with each new paper on the subject.

When the story resumes again, the scene is once more Flagstaff, where the remote orb of Lowell's dreams had never been far from the minds of his loyal staff. Unfortunately, Lowell's widow tied up her husband's estate for years in an expensive and bitter legal struggle, and it was necessary to keep the observatory's operating expenses to a minimum. Specifically, there were no funds to acquire a proper telescope for making a renewed search. At last, Lowell's brother Abbott Lawrence Lowell, then president of Harvard, came to the rescue with a gift of $4,000, enabling the observatory to acquire a 13-inch photographic refractor.

There were at the time only three astronomers on the Lowell staff—V. M. Slipher, the director, his brother E. C. Slipher, and C. O. Lampland. They were busy with their own research and could not spare time for the work of exposing plates in a new planet search. Clearly what was needed was an assistant, and just then a young farmer from Kansas, Clyde W. Tombaugh, wrote the observatory asking for an opinion about some sketches he had made of Mars and Jupiter. He had taught himself all the astronomy he knew, and even his telescope, a 9-inch reflector, was homebuilt. Its base had been part of a cream separator, while another component was the crankshaft of his father's 1910 Buick.

Director Slipher was sufficiently impressed with Tombaugh's sketches to offer him the job. As he wrote, "It seemed to me that we would probably get more real assistance from this young man than we would from the highly trained variety for the reason that the latter care only to take up new pieces of work for themselves rather than help us with lines the Observatory has been doing."

Tombaugh immediately took the train out west, arriving in Arizona in January 1929, a month ahead of the 13-inch telescope. It took several more months to set up the telescope and make the necessary adjustments, but finally it was ready, and on April 6, 1929, Tombaugh exposed his first plate.

Tombaugh's initial instructions from Slipher were to do the regions in Gemini and then proceed east along the ecliptic as rapidly as possible. Gemini was of interest because the Planet X of Lowell's last predictions was supposed to be lurking there in 1929. On April 11 and 30, Tombaugh exposed plates of the region around Delta Geminorum. The first of these broke into two pieces, and there were several other plate casualties involving this field. Tombaugh has said, "I ought to have had the presence of mind to know that this was the place to find the elusive little rascal."

In April, Gemini was already ninety degrees past opposition. Working backward through the zodiac, by mid-June Tombaugh had caught up with the opposition point in the star-rich regions of the Milky Way in Scorpius and Sagittarius. The star images in these plates were, he noted, three times as thick as in those of Gemini; each 14-by-17-inch plate contained upwards of a million stars.

Initially, as we have seen, Lowell scrutinized his plates with a hand magnifying glass, but he later scrapped this highly inefficient method and acquired a blink comparator—a device employing an electromagnet to flip a small mirror back and forth so as to redirect the light path from one plate to the other. In this way it was possible to examine corresponding regions on the plates in rapid succession, with the image of any planet that registered on exposures taken several days apart betraying itself immediately by "jumping" back and forth. This brought about an enormous improvement in the speed and accuracy with which one could sort through the millions of star images, any one of which could be the sought-for planet.

The first Gemini plates had been blinked by the Sliphers. Nothing turned up, however, and as no further blinking was done in the ensuing months, plates began to accumulate. Tombaugh still assumed that the older astronomers would do all of the blinking, as whoever blinked the plates would have the ultimate responsibility for the success or failure of the search. He doubted that it would be entrusted to an assistant. But in mid-June, as Flagstaff's cloudy season approached, V. M. Slipher assigned Tombaugh the task of blinking the plates. Tombaugh suspects that perhaps unconsciously the older men had given up. There was, however, a more transparent motive, for as Tombaugh found, blinking these plates was "the most tedious work I had ever done." Moreover,

> I encountered several dozen asteroids which shifted in position in the interval between the dates the plates were taken. How would I know which one was Planet X? Several of these plate regions had no third plate and some were taken near the asteroid stationary points. I just could not reconcile myself to conducting a planet search on such a hit-and-miss manner. I was in a state of despair.

The asteroid stationary points occur where the asteroids, in changing direction in their retrograde loops, appear to slow down nearly to a standstill and are thus only too apt to be mistaken for a slowly moving outer planet.

Tombaugh pondered the problem of the asteroid stationary points throughout the summer. Finally the solution dawned on him. By exposing his plates only near the opposition point, any body orbiting beyond the Earth would be caught during the time of its most rapid retrograde motion through the field. This was crucial. Remember, the retrograde loops are actually reflections of the Earth's movement produced by parallax, so the nearer the moving body is to the Earth, the faster will be its motion through the loop. Thus, by exposing the plates only near the opposition point, not only would Tombaugh avoid the asteroid stationary points but also the relative distance of any body recorded on the plates would become apparent at once.

As the rainy season ended that fall, Tombaugh, with new enthusiasm, began chasing the opposition point through Aquarius and Pisces. Among his other precautions, he made his exposures only when the opposition point was nearly overhead so as to eliminate distracting displacements of star images due to atmospheric refraction, and he also obtained a third plate of each region whenever possible in order to rule out planet suspects due to chance concentrations of silver grains in the plate emulsion.

By January 1930, Tombaugh was again photographing the Delta Geminorum region, where he had had such difficulty getting a plate the year before. He obtained three plates, on January 21, 23, and 29. Of these, the first was of only marginal quality owing to poor weather conditions, but the others were good. Moonlight began interfering with photography on February 7, and at that point Tombaugh went back to blinking. He finished blinking a pair of plates of the eastern Taurus region, and then, about February 15, he put the Delta Geminorum plates up on the comparator. By late afternoon on February 18 he had blinked about a quarter of the pair. What happened next is best told in his own words:

> I raised the eyepiece assembly to the next horizontal strip. At the center line, I had the guide star Delta Gem in the small rectangular field of the eyepiece. After scanning a few fields to the left, I turned the next field into view. Suddenly I spied a fifteenth magnitude image popping out and disappearing in the rapidly alternating views. Then I spied another image doing the same thing about 3 millimeters to the left. "That's it," I exclaimed to myself.

He verified that the image on the January 23 plate was to the east of that recorded on January 29. This proved that the images were indeed produced by an object in retrograde motion. Moreover, considering the interval between the plates, the parallactic shift indicated that the object was far beyond the orbit of Neptune. His third plate, though of poor quality, showed an image just where it was supposed to be, and he now had no doubt that the planet was in his grasp.

Tombaugh ran down the hall to V. M. Slipher's office and announced: "I've found Planet X. I'll show you the evidence." Slipher was impressed, but he did not announce the discovery immediately. First he wanted to be absolutely sure that what Tombaugh had discovered was trans-Neptunian and a planet. Finally, on March 12, Slipher telegraphed news of the discovery to the Harvard College Observatory, and the following day—the 75th anniversary of Lowell's birth and the 149th of Herschel's discovery of Uranus—word went out to the rest of the world: "Systematic search begun years ago supplementing Lowell's investigation for Trans-Neptunian planet has revealed object which since seven weeks has in rate of motion and path consistently conformed to Trans-Neptunian body at approximate distance he assigned." Lowell was prominent in Slipher's telegram; Tombaugh went unmentioned.

At the time it was assumed throughout the astronomical world that the planet was Lowell's Planet X. After all, it was found only 5.9° from Lowell's predicted longitude, and its orbit bore a "remarkable" similarity to the postulated orbit, a point Slipher underlined at the time.

Lowell had situated his planet at a mean distance from the Sun of 43 AU and had assumed a highly elliptical orbit with an inclination to the ecliptic of 10°. The planet actually lay at a mean distance of 39.4 AU, and its orbit proved to be the most eccentric and sharply inclined of any planet. When farthest from the Sun, as it will next be in 2113, it lies at a distance of 7.4 billion kilometers. At perihelion, which it passed in 1989, its distance is only 4.4 billion kilometers, so that it passes inside the orbit of Neptune— though I hasten to add that there is not the slightest chance of a collision because, owing to the extreme tilt of the planet's orbit (17°), it is actually more than a billion kilometers above the plane of Neptune's orbit at the time. The sharp inclination also means, by the way, that Lowell's 1905–7 search, which was concentrated on the ecliptic, did not have the slightest chance of success, for the planet was so far from the ecliptic at that time that it was not even in the zone covered by the plates.

Since the planet takes 248 years to make each circuit of the Sun, it will not return to the part of its orbit where Tombaugh discovered it until 2178, and for its next perihelion passage we will have to wait until 2237. After the calculation of an orbit, the next most urgent business was to find a suitable name for the planet. Mrs. Lowell at first favored Zeus but later changed her mind and decided that the planet should be named Constance after herself. Neither of these suggestions was much favored by anyone else. Instead, the leading candidates were Cronos, Minerva, and Pluto. Cronos was eliminated when people realized that the name had been given to one of the planets postulated by T.J.J. See, an astronomer whose personality made him universally detested and earned for him epithets like "reptile." The name Minerva had already been bestowed on one of the asteroids, so this left Pluto. The name was singularly apt in any case, as Pluto was the god of the gloomy underworld and it recalled in its first two letters the name of Percival Lowell.

Mrs. Lowell visited the observatory in the summer of 1930. She was still dressed in mourning black, and Tombaugh recalls that she was eager to meet the young man who had found "my husband's planet." But already doubts were beginning to surface about the appropriateness of the possessive. First, there was a rival claimant: Pickering, who was quick to point out that whereas Pluto was 5.9° from Lowell's predicted position, it was only 5.6° from where his Planet O of 1919 was supposed to be lurking. Indeed, Pluto had actually been registered on the plates Humason had taken at Mt. Wilson in 1919, but it had been overlooked because in one plate its image had fallen on a defect and in the other it had been swamped by glare from a nearby star. Pickering neglected to note, of course, that he had published a revised orbit for Planet O in 1928 which was much poorer, and though the Lowell Observatory seems to have fretted about Pickering for a while, probably the final verdict on him was accurately pronounced by Lowell Observatory trustee Roger Lowell Putnam, Percival's nephew, who wrote to V. M. Slipher: "I don't think I should worry much about Pickering's predictions. I think he has predicted just about everything—from one planet to three, in varying positions. At any rate, Dr. Lowell predicted it, and you have found it, which is more than Pickering has done."

A weightier challenge came from Yale celestial mechanician Ernest W. Brown, who announced that Pluto could not possibly be the planet of Lowell's prediction despite the impressive similarity in the orbits. Brown's main point was that Pluto was simply not massive enough to have caused the perturbations in Uranus's motion that Lowell had assumed. Even in large telescopes it was impossible to make out a disk—in the Lowell 24-inch refractor the upper limit was put at 0.5 seconds of arc. This meant that even with the most generous assumptions (that the planet was made up entirely of iron, for instance) its mass still came to at most a third of that required by Lowell.

Not everyone was pleased with this conclusion, least of all the astronomers at the Lowell Observatory, and it was difficult to accept that Lowell's seemingly uncanny prediction could have been the result of mere chance. Over the years, various attempts were made to square the prediction with Pluto's apparent small size. According to one theory, for example, the planet's ice-covered surface acted like a curved mirror to form a virtual image of the Sun, and it was that, not Pluto's actual disk, that was being measured in the telescope. Others were willing to believe that Pluto was as small as it seemed but wondered whether it deserved the status of a proper planet at all. It was even suggested that perhaps Pluto had once been a satellite of Neptune and that on veering too close to Triton it had been ejected into its highly unusual orbit, while at the same time the direction of Triton's motion had been reversed from direct to retrograde.

As Pluto drew steadily nearer the Earth as it approached its perihelion passage of the Sun, its diameter and mass were revised steadily downward. By 1976 the best estimate of its diameter had fallen below 3,000 kilometers, and of its mass to a few thousandths that of the Earth. Then, in 1978, James W. Christy of the U.S. Naval Observatory discovered Pluto's satellite, Charon, named for the grim ferryman of the underworld. It revolves around Pluto once every 6 days, 9 hours, the same period as that in which the planet rotates, which means, by the way, that it must always remain stationary over the same part of the planet. The discovery of a satellite allowed Pluto's mass to be calculated directly from Kepler's third law. The result was even lower than expected: Pluto had only a fifth of the mass of the Earth's moon, or about two thousandths that of the Earth. Clearly it was too small to have reversed Triton's motion, and it was too small to have disturbed Uranus's motion to the extent assumed by Lowell. It could not have been Lowell's Planet X; this was certain.

The discovery of Charon could not have been more timely, as it came on the eve of a rare series of mutual events between the satellite and its primary. As seen from the Earth, Pluto and Charon played hide-and-seek behind one another between 1985 and 1990, a series of events observable only once every 124 years. Largely from observations of these mutual events, a fair amount is now known about the Pluto-Charon system. The distance between the centers of the two bodies is 19,640 kilometers, and Pluto's diameter, so long uncertain, has now been accurately measured as 2,302 kilometers. Charon is 1,186 kilometers across, which is 52 percent of the diameter of Pluto, so that relative to its primary it is the largest satellite in the Solar System—an honor formerly held by the Earth's moon.

Pluto and Charon consist largely of rocky materials, unlike many of the icy outer-planet satellites visited by Voyager 2—a significant exception being Triton, which has a similar composition. Another oddity is that Pluto has a highly tilted axis of rotation. In this respect it resembles Uranus.

Careful observations of changes in light intensity during Charon's passages in front of Pluto have revealed that Pluto has bright polar caps and a slightly reddish equatorial zone. Charon, in contrast, is dark gray.

Since 1976, Pluto's surface has been known to contain methane frost, and no doubt this is the composition of its polar caps. When fresh, methane frost is bright, which is consistent with the caps' high reflectivity. However, such a covering would soon be darkened by the conversion of methane to more complex hydrocarbons unless it were continually renewed.

For a few years around the time of Pluto's perihelion passage, the Sun warms its surface to just over $-190°C$, the critical temperature at which methane sublimes, or changes from a solid to a gas. As the methane frost sublimes away, more and more of the dark, hydrocarbon-rich surface below is exposed. Consistent with this, a steady darkening of the planet was observed during the decades leading up to its perihelion passage in 1989. The methane forms a short-lived midsummer atmosphere the pressure of which reaches perhaps 1/100,000th that of the Earth at sea level—a low figure, admittedly, but significant. For the next decade or two, as the planet withdraws from the Sun, this methane atmosphere freezes out again, covering the planet with fresh methane "snow" and causing a corresponding increase in brightness. Clearly, Pluto is a world of strange seasons, and its surface must be just as active as that of Triton, which from close range it will probably be found to resemble. Charon, too, may well have a transient atmosphere around the time of its perihelion passage.

Where did Pluto and Charon come from? They give every indication of being wrenched from another part of the Solar System, fragmentary products of some primordial catastrophe. Perhaps the old idea that Pluto was formerly a satellite of Neptune ought to be revived, though in a modified form. Certainly the fact that Pluto and Charon cross the orbit of Neptune is highly suspicious—it is only too appealing to think that they here return to the scene of their birth. Moreover, Triton, with its rocky core, sharply tilted axis, and steeply inclined orbit, around which it perversely travels in a retrograde direction, seems in all respects a cousin to these peculiar objects. It may well be a remnant of the same disaster.

A planet several times as massive as the Earth and rushing through the Neptunian system could conceivably have produced this set of anomalous objects, turning Triton around in its orbit, breaking up another satellite and flinging it and its largest fragment into bona fide planetary paths, only to vanish from the scene. But if this wreckage—Triton, Pluto, and Charon—are not themselves all that remains of that intruder, it may be that the rest of it will someday turn up, wandering in a highly elliptical and sharply inclined orbit. This brings us to the next question: Is the Solar System as we now know it complete, or is there perhaps a planet (or planets) remaining to be discovered?

While publicly identifying Pluto with Lowell's Planet X, within a year of Pluto's discovery Lowell Observatory astronomers began to fear that the planet Lowell had predicted remained somewhere out there, hidden in the camouflage of stars. Clyde Tombaugh was assigned to carry on the search with the same methods that had been so successful in bringing Pluto to light. Over the next thirteen years he covered 70 percent of the sky down to stars of the sixteenth and seventeenth magnitudes, including areas far from the ecliptic where a planet with a highly inclined orbit might be lurking. He found no new planet. Knowing Tombaugh's thoroughness, it is hard to imagine that he could have missed anything. Yet certain residuals of both Uranus and Neptune remain and need to be explained, and several astronomers have gone so far as to calculate the position of a postulated perturber, though admittedly their results are not in close agreement.

One can at least say rather definitely where the planet cannot be. Since no deflections have been noted in the paths the two Pioneer spacecraft followed through the outer Solar System (in nearly opposite directions, be it noted) it cannot have been anywhere close to them. Moreover, Voyager 2 found Uranus and Neptune right where they were supposed to be, based on orbits calculated from observations made only since 1910. One's confidence in the existence of a Planet X therefore boils down, finally, to one's faith in the pre-1910 observations. If the old observations are indeed accurate, one can say that Planet X was in a position to disturb Uranus and Neptune before and during the nineteenth century but has since moved outward in such a highly elliptical and inclined orbit that its effects are no longer perceptible and indeed may not be again for centuries.

Certainly those who doubt the existence of Planet X may carry the day, but at least one prediscovery observation is especially suggestive. Taken at face value, it indicates that in January 1613 Neptune was a full minute of arc away from where it ought to have been. There is only one man who could have made this observation: Galileo himself, who at the time was mapping the motion of Jupiter and its moons among the stars. On January 27 he noted an eighth-magnitude star roughly in line with Jupiter, its moons, and another star. When observing again the following night, he commented that the two stars "seemed farther apart" and even made a sketch, but he failed to pursue the matter further. The eighth-magnitude star, we now know, was Neptune.

Of course, it may well be that Galileo simply erred in the position in which he indicated his star, but it is also possible that his observation indicates the presence of a disturber in that part of Neptune's orbit. If so, it may eventually turn up—and at the moment, the best chance is that it will be captured during Eleanor Helin's survey of the outer Solar System using the 18-inch Schmidt telescope at Mount Palomar. Unfortunately, calculated positions are not apt to be of much help, at least for the foreseeable future. The discovery will fall only to dedicated searching.

Planet-searching has become much more difficult than in the days of Galle and d'Arrest, who found Neptune in an hour's work at the telescope. Any unknown planet will have to be distinguished from scores of millions of stars, the heartbreaking work of years—without, of course, the slightest guarantee that a planet remains to be discovered.

CHAPTER FOURTEEN

Comets and Meteors

During the years 1979 to 1999, Neptune will be the outermost known planet. Thereafter Pluto will reassume the mantle. Well out beyond either lies the heliopause, where the solar wind collides with the interstellar medium, and beyond that, from perhaps 10,000 or 20,000 AU out to 200,000 AU from the Sun, one encounters a circular shell of planetesimals left over from the primeval nebula—it is known as the Oort Cloud, after the Dutch astronomer Jan Oort, who first described it in 1950.

These planetesimals are the comets, and every now and then one of them is jarred loose and starts on its long sunward journey. If it should pass too close to one of the giant planets, it may be captured and become a short-period comet, of which Halley's is the best-known example— its aphelion lies out beyond Neptune, but about once every seventy-six years it swings through the inner Solar System. The giant planets may exert a slingshot effect on some comets and accelerate them out of the Solar System altogether. Many comets remain in closed if highly elongated paths that take them back from whence they came, not due to return again to the Sun's vicinity for thousands or even millions of years.

In structure, a comet consists of the nucleus, the coma, and if it is large enough, one or more tails. The nucleus is the comet's core; it is made up mostly of water ice and dust, but it contains other substances as well, including hydrogen cyanide, carbon monoxide, and carbon dioxide, and it comprises most of the comet's mass. So far the nucleus of Halley's comet has been the only one imaged at close range. In the photographs taken by the European spacecraft Giotto in 1986, it appears as a coal-black potato-shaped body some 16 kilometers long by 8 kilometers wide. Its surface is uneven, and various irregularities have been described as "mountains" and "craters," though their true nature is not yet known.

Telescopically, the nucleus itself cannot actually be seen, because it is hidden by the luminous gas and dust of the coma, the central condensation of which may appear as a brilliant starlike point, the "false nucleus." Sometimes a series of hoods may surround the nucleus. This was true of Donati's comet of 1859 and of the Great Comet of 1874, the hoods of which expanded from hour to hour.

Jets and fountains, erupting from the nucleus as icy materials vaporize in the intensifying sunlight, are typically seen in comets that pass within half an astronomical unit of the Earth. Swept back by the Sun, the streaming gases may appear as straight rays or may spiral into intricate catherine-wheel structures as the nucleus rotates. Such structures are best studied visually. In long-exposure photographs such delicate features tend to be lost in the overall brilliance of the coma. Interestingly, cometary jets act like the thrusters of spacecraft, producing small deflections in the comet's motion. This is why comets are frequently unpunctual visitors—Halley's in 1910, for instance, appeared a full three days behind schedule even though all planetary perturbations were taken into account.

It is their tails which make comets such awe-inspiring sights. They are of two main types. Gas or plasma tails, also known as Type I, are thin and straight, and appear bluish because of light emitted by carbon monoxide ions. Dust tails, or Type II, are broad, yellow, and curving. Needless to say, dust tails develop only if the ices of the nucleus contain significant amounts of dust, and a given comet may well sport tails of both types—or none at all, as is the case of many of the smaller ones, which happens to include most of those having short periods.

Halley's comet is so well known both because it is a fairly large comet and because the period between its returns is reasonably short. Its orbit was first worked out by Edmund Halley, who, upon reviewing the records of all the comets that had appeared between 1337 and the end of the seventeenth century, discovered that the comets that appeared in 1531, 1607, and 1682 all followed similar paths. He concluded that they were one and the same and predicted another return in 1758. Halley died in 1742, but the comet was duly recovered on Christmas Day 1758 by a German amateur, Johann Palitzsch, and it returned again in 1835, 1910, and 1986. It is next due back in 2061. Interestingly, it travels around its elongated orbit in a retrograde direction.

Among other notable comets, Chéseaux' comet of 1744 had seven tails, and Lexell's of 1770 passed within only 2 million kilometers of the Earth and had a head whose apparent diameter was five times the breadth of the full moon. The Great Comets of 1811, 1843, and 1861 had tails stretching 100° or more across the sky, while the dust tail of Donati's comet of 1858, which is generally considered the most beautiful comet of the nineteenth century, was gracefully curved, like a scimitar. The Great Comet of 1882 and the "Daylight Comet" of 1910 (not to be confused with Halley's comet, which appeared later that same year) were bright enough to be seen in broad daylight.

During the nineteenth century a "great" comet appeared about once every decade. The twentieth century has not been so favored, yet there have been several notable comets in recent years. Comet Ikeya-Seki swept to within 11 million kilometers of the Sun in October 1965, and as it swept back out into the depths of space it developed a spectacularly long tail (fig. 14.1). I well remember its tail, which appeared as a ghostly plume climbing the heavens before sunrise. It was easily visible well before the comet's head cleared the horizon. Ikeya-Seki will next return in about eight hundred years. Comet Bennett put on a brave show for Northern Hemisphere observers in the predawn hours in the spring of 1970 and was remarkable for the spiral form of its jets. Even more spectacular was Comet West of 1976 (fig. 14.2), which had a well-developed dust tail that fanned out into an array of beautiful streamers, making it the most aesthetically pleasing of recent memory.

A good example of comets' notorious unpredictability was Comet Kohoutek of 1973. It was the subject of much advance publicity because of its brilliance while still at an enormous distance from the Sun, and it did make a rewarding object in binoculars. But it failed to live up to expectations, fizzling badly as it approached the Sun. Halley's comet was also a disappointment, though for entirely different reasons, during its most recent return in 1986. It missed the Earth by almost the widest distance possible, passing perihelion when on the side of the Sun opposite the Earth, and its closest approach was 62 million kilometers, or farther than Mars at its closest.

Of course, a brilliant new comet may appear at any time, and as it is the custom to bestow on a comet the name of its discoverer, a certain immortality may be attained thereby. As the Japanese comet hunter Karou Ikeya put it, finding a comet offers a chance to "write my name across the sky." Ikeya found six, including a share in the discovery of the twentieth century's brightest comet, Ikeya-Seki.

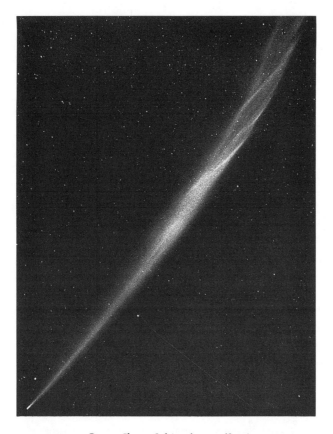

FIGURE 14.1. Comet Ikeya-Seki, whose official name is
1965 VIII, signifying that it was the eighth comet to pass
perihelion that year. Ikeya-Seki was one of the most
spectacular comets of the twentieth century. It was briefly
visible in broad daylight after its perihelion passage in
October 1965. (Courtesy Lick Observatory)

Love of fame motivated Charles Messier of France, who became the
leading discoverer of the eighteenth century, with twenty-one to his credit.
His record was surpassed by Jean-Louis Pons, who began his career as the
concierge at the Marseilles Observatory but who went on to become one
of the world's leading astronomers. Pons had discovered thirty-seven
comets by the time he died in 1831. Among other leading comet discoverers, W. R. Brooks and E. E. Barnard were active at the end of the last
century and were credited with twenty-seven and nineteen comets, respectively. Among those living today, Carolyn Shoemaker leads with twenty-

FIGURE 14.2. Comet West (1976 VI). The gas
tail extends straight back from the head, and the
several curving dust tails have a windswept
appearance. (Courtesy Lick Observatory)

two, followed by William Bradfield with thirteen, while the most successful amateur discoverer of the past few years has been David Levy, who so
far has discovered sixteen. One of them, which he found in May 1990,
brightened to naked-eye visibility late that summer.

Since part of the comets' material evaporates each time they come to
perihelion, they cannot last forever, and those with the shortest periods
may have lifetimes lasting only a few centuries. There is no real cause for
concern, however, as the supply is continually being replenished from the
Oort Cloud, which some estimate may contain upwards of a trillion
potential comets. Some comets have actually been observed undergoing
disintegration. This was true of Comet West, for example, which fragmented into four parts as it rounded the Sun in 1976. Each of these
fragments will itself return as a separate comet about a half million years
hence.

What becomes of comets that have disintegrated? The debris they have strewn along their paths continues to round the Sun as meteoroid streams, and if the Earth, in pursuing its own orbit around the Sun, passes through the thickest part of such a stream, the result is a meteor shower. The meteors are tiny particles vaporizing as they enter the atmosphere at high speeds.

Usually the best annual display is put on by the Perseids in August, which consist of debris left behind by Comet Swift-Tuttle (meteor showers receive their name from the constellation from which the meteors appear to radiate, in this case Perseus). The November Leonids, which follow the orbit of Comet Tempel-Tuttle, put on displays in 1833 and 1866, were missed in 1899 and 1933 (we now know that planetary perturbations had pulled them out of their former path) and returned in full force in 1966. In Minnesota I saw about a thousand in a period of an hour, while in the western United States the shower was even heavier. There is every reason to expect another brilliant display in 1999. Two annual showers are associated with Halley's Comet, the Eta Aquarids in May and the Orionids in October. Most of the particles scattered by comets along their paths are very tiny, and they burn up completely in the Earth's atmosphere. Meteorites, which are chunks of material large enough to reach the ground, for the most part derive not from comets but from asteroids that have been ejected from the main belt. In a few cases, meteorites have been identified as having come from the Moon or Mars.

What I have said about comets and meteors is clearly only the briefest of introductions, but I do hope it conveys some sense of their importance. Comets have brought about the deaths of worlds—as earlier described, their collisions with small moons are believed to have produced the planetary rings—and a cometary or asteroidal impact has been strongly implicated in the dinosaurs' demise. Insofar as this is true, they doubtless have their sinister aspect, and this is how superstition has always portrayed them. But there is a prologue as well as an epilogue to their story: they have presided over births as well as deaths, for it was from planetesimals such as these that the planets themselves formed during the violent, heroic age of the Solar System's creation 4.6 billion years ago.

Epilogue

When the Voyager spacecraft swept by Neptune in August 1989, an era came to an end. The first spacecraft reconnaissance of the Solar System was complete. No doubt other missions will follow, adding significant new discoveries. Even as I write, the Magellan spacecraft is radar mapping the surface of Venus. But it is doubtful that we will ever again experience the exhilaration afforded by these first close-up views, when startling vistas of planets, rings, and moons flooded upon us in dizzying profusion.

Perhaps even more important than the breathtaking views we have had of other worlds—and brave new worlds many of them have proved to be—is the fresh perspective we have gained on our own world. From orbiting spacecraft in the early 1960s we first saw the Earth as a globe—the gray, brown, and green stretches of its lands, the blue vastness of its oceans, all encompassed within a thin layer of air. In December 1968 we saw it from the Moon, an oasis of blue and white amid the darkness. Then in September 1977 the crescents of the Earth and Moon were framed together for the first time as Voyager 1 looked back at them from a distance of 12 million kilometers as it hurtled toward the outer Solar System. Each image brought a new vision of the Earth.

Finally we received the latest in this remarkable series, captured in February 1990 when Voyager 1 was 6 billion kilometers away. The result, though visually less spectacular than the others, was no less momentous. Indeed, its importance lay in its very lack of spectacle. From this great distance, the Earth filled less than a single picture element, or pixel. There it was: a spot, a smidgen of pale blue surrounded by an immense darkness, an utterly insignificant place, our home.

APPENDIXES

Appendix 1. Statistical Information Concerning the Major Planets

Planet	Mean Distance from the Sun		Eccentricity of Orbit	Period of Revolution		Inclination of Orbit to Ecliptic
	Millions of mi	Millions of km		Sidereal	Synodic	
Mercury	36.0	57.9	.206	88.0 d	116 d	7.0°
Venus	67.2	108.1	.007	224.7 d	584 d	3.4
Earth	92.9	149.5	.017	365.26 d	—	0.0
Mars	141.5	227.9	.093	687.0 d	780 d	1.8
Jupiter	483.3	778.3	.048	11.86 yr	339 d	1.3
Saturn	886.2	1,427.0	.056	29.46 yr	378 d	2.5
Uranus	1,783.1	2,869.6	.047	84.01 yr	370 d	0.8
Neptune	2,794.0	4,496.6	.009	168.79 yr	367 d	1.8
Pluto	3,666.0	5,900.0	.250	247.69 yr	367 d	17.2

[a]Retrograde
[b]Parentheses indicate an approximate value.

Equatorial Diameter		Oblate-	Mass	Density	Escape Velocity			Inclination of Equator to
miles	kilometers	ness	(Earth = 1)	(gm/cm³)	mi/sec	km/sec	Rotational Period	Plane of Orbit
3,031	4,878	0	0.0553	5.43	2.7	4.3	58.64 d	2
7,521	12,104	0	0.8150	5.24	6.5	10.4	243 d[a]	178
7,926	12,756	1/298	1.0	5.52	7.0	11.2	23 hr 56 min 4.1 sec	23.4
4,217	6,787	1/193	0.1074	3.94	3.1	5.0	24 hr 37 min 22.3 sec	25.2
88,730	142,800	1/15	317.83	1.33	37.0	59.6	9 hr 50 min [equatorial] 9 hr 55 min [polar]	3.1
74,900	120,540	1/9	95.18	0.70	22.1	35.5	10 hr 39 min	26.7
31,763	51,118	1/45	14.53	1.29	13.26	21.3	17 hr 54 min	97.9
30,775	49,528	1/40	17.13	1.64	14.48	23.3	19 hr 12 min	29.6
1,430	2,300	(0)[b]	0.002	2.03	0.68	1.1	6 d 9 hr	122.5

Appendix 2. Satellites of the Solar System

Planet	Satellite	Discoverer and Year of Discovery	Visual Magnitude at Mean Opposition	Diameter (in km)	Mean Distance from Planet's Center (in km)
Earth	Moon	—	−12.7	3,476	384,400
Mars	Phobos	A. Hall, 1877	12	27 × 19	9,400
	Deimos	A. Hall, 1877	13	15 × 11	23,500
Jupiter	Metis	Voyager 2, 1980		(40)[b]	127,960
	Adrastea	Voyager 2, 1979		24 × 16	128,980
	Amalthea	E. Barnard, 1892	14	270 × 150	181,300
	Thebe	Voyager 2, 1979		(100)	221,900
	Io	Galileo, 1610	5.0	3,630	421,600
	Europa	Galileo, 1610	5.3	3,138	670,900
	Ganymede	Galileo, 1610	4.6	5,262	1,070,000
	Callisto	Galileo, 1610	5.6	4,800	1,883,000
	Leda	C. Kowal, 1974		(16)	11,094,000
	Himalia	C. Perrine, 1904		(180)	11,480,000
	Lysithea	S. Nicholson, 1938		(40)	11,720,000
	Elara	C. Perrine, 1905		(80)	11,737,000
	Ananke	S. Nicholson, 1951		(30)	21,200,000
	Carme	S. Nicholson, 1938		(45)	22,600,000
	Pasiphae	P. Melotte, 1908		(70)	23,500,000
	Sinope	S. Nicholson, 1914		(40)	23,700,000
Saturn	Pan	Voyager 2, 1990		(10)	133,583
	Atlas	Voyager 1, 1980		40 × 30	137,640
	Prometheus	Voyager 1, 1980		140 × 80	139,350
	Pandora	Voyager 1, 1980		110 × 70	141,700
	Janus	A. Dollfus, 1966		220 × 160	151,450
	Epimetheus	J. Fountain and S. Larson, 1978		140 × 100	151,450
	Mimas	W. Herschel, 1789	13	390	185,520
	Enceladus	W. Herschel, 1789	12	500	238,020
	Tethys	G. Cassini, 1684	10.3	1,050	294,660
	Telesto	Voyager 1, 1980		(25)	294,660
	Calypso	Voyager 1, 1980		30 × 20	294,660
	Dione	G. Cassini, 1684	10.4	1,120	377,400
	Helene	P. Laques and J. Lecacheux, 1980		36 × 30	377,400
	Rhea	G. Cassini, 1672	9.7	1,530	527,040
	Titan	C. Huygens, 1655	8.4	5,150	1,221,850
	Hyperion	G. Bond, 1848	14	350 × 200	1,481,000
	Iapetus	G. Cassini, 1671	11–13	1,440	3,561,300
	Phoebe	W. Pickering, 1898		220	12,952,000
Uranus	Cordelia	Voyager 2, 1986		(30)	49,750
	Ophelia	Voyager 2, 1986		(30)	53,760
	Bianca	Voyager 2, 1986		(40)	59,160
	Cressida	Voyager 2, 1986		(70)	61,770
	Desdemona	Voyager 2, 1986		(60)	62,660
	Juliet	Voyager 2, 1986		(80)	64,360

Planet	Satellite	Period of Revolution d hr min	Orbital Eccentricity	Inclination	Mass (Earth's moon = 1)[a]
Earth	Moon	27 07 44	0.055	18–29°	1.00
Mars	Phobos	00 07 39	0.015	1.1	1.8×10^{-7}
	Deimos	01 06 18	0.000	1.8	2.4×10^{-8}
Jupiter	Metis	00 07 05	0.000	0.0	
	Adrastea	00 07 09	(0)[b]	0.0	
	Amalthea	00 11 57	0.00	0.4	
	Thebe	00 16 12	0.01	0.0	
	Io	01 18 27	0.00	0.0	1.21
	Europa	03 13 13	0.01	0.5	0.66
	Ganymede	07 03 43	0.00	0.2	2.03
	Callisto	16 16 32	0.01	0.2	1.46
	Leda	238 17 17	0.15	27	
	Himalia	250 13 41	0.16	28	
	Lysithea	259 05 17	0.11	29	
	Elara	259 15 36	0.21	28	
	Ananke	631 00 00[c]	0.17	147	
	Carme	692 00 00[c]	0.21	163	
	Pasiphae	735 00 00[c]	0.38	147	
	Sinope	758 00 00[c]	0.28	153	
Saturn	Pan	00 11 48			
	Atlas	00 14 27	(0)	0.3	
	Prometheus	00 14 43	0.00	0.0	
	Pandora	00 15 06	0.00	0.1	
	Janus	00 16 41	0.01	0.3	
	Epimetheus	00 16 39	0.01	0.1	
	Mimas	00 22 36	0.02	1.5	5.2×10^{-4}
	Enceladus	01 08 53	0.00	0.02	(0.001)[b]
	Tethys	01 21 19	0.00	1.1	0.010
	Telesto	01 21 19	(0)	(0)[b]	
	Calypso	01 21 19	(0)	(0)	
	Dione	02 17 41	0.00	0.02	0.014
	Helene	02 17 41	0.01	0.2	
	Rhea	04 12 26	0.00	0.4	0.034
	Titan	15 22 41	0.03	0.3	1.83
	Hyperion	21 06 39	0.10	0.4	
	Iapetus	79 07 57	0.03	14.7	0.026
	Phoebe	550 11 31[c]	0.16	150	
Uranus	Cordelia	00 08 02	(0)	(0.1)	
	Ophelia	00 09 01	(0.01)	(0.1)	
	Bianca	00 10 26	(0)	(0.2)	
	Cressida	00 11 08	(0)	(0.0)	
	Desdemona	00 11 23	(0)[b]	(0.2)[b]	
	Juliet	00 11 50	(0)	(0.1)	

Appendix 2. *Continued*

Planet	Satellite	Discoverer and Year of Discovery	Visual Magnitude at Mean Opposition	Diameter (in km)	Mean Distance from Planet's Center (in km)
	Portia	Voyager 2, 1986		(110)[b]	66,100
	Rosalind	Voyager 2, 1986		(60)	69,930
	Belinda	Voyager 2, 1986		(70)	75,260
	Puck	Voyager 2, 1985		150	86,010
	Miranda	G. Kuiper, 1948		470	129,780
	Ariel	W. Lassell, 1851	14.4	1,160	191,240
	Umbriel	W. Lassell, 1851	15.3	1,170	265,970
	Titania	W. Herschel, 1787	14	1,580	435,840
	Oberon	W. Herschel, 1787	14.2	1,520	582,600
Neptune	Naiad	Voyager 2, 1989		(50)	48,000
	Thalassa	Voyager 2, 1989		(80)	50,000
	Despoina	Voyager 2, 1989		(180)	52,500
	Galatea	Voyager 2, 1989		(150)	62,000
	Larissa	Voyager 2, 1989		(190)	73,600
	Proteus	Voyager 2, 1989		(400)	117,600
	Triton	W. Lassell, 1846	14	2,700	354,800
	Nereid	G. Kuiper, 1949		(340)	5,513,400
Pluto	Charon	J. Christy, 1978		1,190	19,640

[a]The actual mass of the Earth's moon is 735×10^{20} kg.
[b]Parentheses indicate an approximate value.
[c]retrograde

Planet	Satellite	Period of Revolution d hr min	Orbital Eccentricity	Inclination	Mass (Earth's moon = 1)[a]
	Portia	00 12 19	(0)	(0.1)	
	Rosalind	00 13 24	(0)	(0.3)	
	Belinda	00 14 59	(0)	(0.0)	
	Puck	00 18 17	(0)	(0.3)	
	Miranda	01 09 56	0.00	3.4	0.001
	Ariel	02 12 29	0.00	0	0.018
	Umbriel	04 09 52	0.00	0	0.017
	Titania	08 16 57	0.00	0	0.047
	Oberon	13 11 07	0.00	0	0.040
Neptune	Naiad	00 07 06	(0)	(4.5)	
	Thalassa	00 07 29	(0)	(0)	
	Despoina	00 07 60	(0)	(0)	
	Galatea	00 10 18	(0)	(0)	
	Larissa	00 13 18	(0)	(0)	
	Proteus	01 02 54	(0)	(0)	
	Triton	05 21 03[c]	0.00	157	0.293
	Nereid	360 03 50	0.75	29	
Pluto	Charon	06 09 17	0.00	(0)	0.03

NOTES

ACKNOWLEDGMENTS

ix "imparts a sense . . . know": H. P. Wilkins, *Our Moon* (London: Frederick Muller, 1954), pp. 173–174.

CHAPTER 1. WANDERING STARS

1 "such irregularity . . . journeys": Sir Thomas Heath, *Aristarchus of Samos: The Ancient Copernicus* (1913; reprint, New York: Dover, 1981), p. 269.

4 "a crank machine . . . advice": Thomas Carlyle, *History of Frederick the Great,* bk. 2, chap. 7.

4 "I wonder . . . will be": Giorgio Vasari, "Filippo Brunelleschi," in *The Lives of the Artists: A Selection,* trans. George Bull (Harmondsworth, Eng.: Penguin, 1965), p. 143.

5 "Technically, perspective . . . point of view": William M. Ivins, Jr., *Art and Geometry: A Study in Space Intuitions* (1946; reprint, New York: Dover, 1964), p. 32.

5 "This happens . . . neutralize one another": Nicholas Copernicus, *Three Copernican Treatises,* trans. Edward Rosen (1939; reprint, New York: Dover, 1959), pp. 77–78.

7 "I awoke . . . on me": J.L.E. Dreyer, *A History of Astronomy from Thales to Kepler* (New York: Dover, 1953), p. 391.

7 "Either the . . . Sun": quoted in Daniel J. Boorstin, *The Discoverers* (New York: Random House, 1983), p. 311.

7 "the celestial . . . weight": quoted in Boorstin, *The Discoverers,* p. 311.

8 "I began to think . . . pretty nearly": quoted in Richard S. Westfall, *Never at Rest: A Biography of Isaac Newton* (Cambridge: Cambridge University Press, 1980), p. 143.

9 "cause . . . geometry": quoted in Richard S. Westfall, *Science and Religion in*

Seventeenth Century England (New Haven: Yale University Press, 1958), p. 196.

CHAPTER 2. THROUGH THE TELESCOPE

10   "Like you . . . I shall forbear": quoted in Giorgio de Santillana, *The Crime of Galileo* (Chicago: University of Chicago Press, 1955), p. 7.

11   "Visible objects . . . nearby": Galileo Galilei, *Discoveries and Opinions of Galileo,* trans. and ed. Stillman Drake (Garden City, N.Y.: Doubleday, 1957), p. 29.

11   "is not . . . valleys": Galileo Galilei, *Discoveries and Opinions,* p. 31.

11   "four wandering stars . . . us": Galileo Galilei, "The Starry Messenger," in *Discoveries and Opinions of Galileo,* trans. Stillman Drake (New York: Doubleday, 1957), p. 28.

12   "vapors . . . fumes": Galileo Galilei, "Letters on Sunspots," in *Discoveries and Opinions of Galileo,* p. 100.

12   "This is . . . suppress": Arthur Koestler, *The Sleepwalkers* (New York: Grosset and Dunlap, 1963), p. 430.

12   "Who should set . . . science": quoted in Santillana, *Crime of Galileo,* p. 39.

15   "Perrault eloquently . . . common sense": quoted in Willy Ley, *Watchers of the Skies* (New York: Viking Press, 1963), p. 169.

16   "Seeing therefore . . . concave metal": Isaac Newton, *Opticks; or, A Treatise of the Reflections, Refractions, Inflections & Colours of Light,* Bk. 1, Pt. I, Prop. VII, Theory VI (New York: Dover, 1982), p. 102.

17   "If the theory . . . a perpetual tremor": Newton, *Opticks,* p. 110.

18   "Against hope . . . my way": Percival Lowell, "Mars," unpublished manuscript in Lowell Observatory Archives.

CHAPTER 3. ROCKETS INTO SPACE

19   "I write of . . . believe them": quoted in Ley, *Watchers,* p. 481.

20   "For a long . . . work of the mind": quoted in John Noble Wilford, *We Reach the Moon* (New York: Bantam, 1969), p. 37.

CHAPTER 4. THE MOON

32   "If you keep . . . solar body": Leonardo da Vinci, *The Notebooks of Leonardo da Vinci,* comp. and ed. Jean Paul Richter (New York: Dover, 1970), 2:167.

35   "Oh, Schroeter . . . in vain": Franz von Paula Gruithuisen, "Entdeckung vieler deutlichen Spuren der Mondbewohner, besonders eines colossalen Kunstgebäudes derselben," *Archiv für die gesammte Naturlehre* (Nuremberg: Johann Leonhard Schrag, 1824), vol. 1, p. 163.

37   "The Moon is . . . Earth": Wilhelm Beer and Johann Heinrich von Mädler, *Der Mond nach seinen kosmischen und individuellen Verhältnissen oder*

Allgemeine vergleichende Selenographie (Berlin: Simon Schropp, 1837), p. 134.

37 "I could not . . . within me": *Iliad*, trans. Richmond Lattimore (Chicago: University of Chicago Press, 1951), Bk. II, lines 488–490.

39 "ninety nine times . . . crest-line," quoted in William Graves Hoyt, *Coon Mountain Controversies: Meteor Crater and the Development of Impact Theory* (Tucson: University of Arizona Press, 1987), p. 58.

40 "seeing that a shot . . . roving masses": quoted in Hoyt, *Coon Mountain Controversies*, p. 69.

44 "One cannot fail . . . side to side": quoted in B. M. Middlehurst and G. P. Kuiper, eds., *The Moon, Meteorites and Comets* (Chicago: University of Chicago Press, 1963), p. 32.

CHAPTER 5. MERCURY

53 "if anyone . . . could be given": quoted in Ley, *Watchers*, p. 192.

55 "Schiaparelli had remarked . . . Australia": E. M. Antoniadi, *Journal of the British Astronomical Association* 45 (1935): 235–236.

62 "an unparalleled triumph . . . within itself": Richard Baum, "Leverrier and the Lost Planet," *Yearbook of Astronomy 1982*, ed. Patrick Moore (New York: W. W. Norton, 1981), p. 152.

63 "my old . . . miseries": Camille Flammarion, *Popular Astronomy*, trans. J. Ellard Gore (New York: Appleton, 1907), p. 347, a translation of an 1880 French edition.

63 "It is you . . . what you have seen": quoted in Baum, "Leverrier," p. 155.

64 "My rough drafts! . . . planing": quoted in Baum, "Leverrier," p. 157.

64 "there has not . . . assassin!": Flammarion, *Popular Astronomy*, p. 349.

65 "Imagine my joy . . . Mercury": letter to Paul Ehrenfest, January 17, 1916, in Abraham Pais, *"Subtle is the Lord": The Science and Life of Albert Einstein* (Oxford: Oxford University Press, 1982), pp. 253–255.

CHAPTER 6. VENUS

67 "It does not seem . . . to the Moon": Antonie Pannekoek, *A History of Astronomy* (1961; reprint, New York: Dover, 1989), p. 35.

69 "could not . . . me": J. H. Schroeter, "New Observations on Further Proof of the Mountainous Inequalities, Rotations, Atmosphere and Twilight of the Planet Venus," *Philosophical Transactions of the Royal Society* 85 (1795), p. 117.

69 "like a confused mass . . . the sunshine," quoted in Richard Baum, *The Planets: Some Myths and Realities* (Newton Abbot: David & Charles, 1973), p. 73.

69 "Observations of . . . Pyrenees": Flammarion, *Popular Astronomy*, p. 369.

71 "Of what nature . . . people our planet," Flammarion, *Popular Astronomy,*
p. 371.

CHAPTER 7. MARS

84 "the inhabitants . . . own": William Herschel, "On the Remarkable Appearances at the Polar Regions of Mars," *Philosophical Transactions of the
Royal Society* 74 (1784), p. 260.

84 "Everything is . . . Earth": quoted in T. W. Webb, *Celestial Objects for
Common Telescopes* (1917; reprint, New York: Dover, 1962), 1:210.

85 "I began to examine . . . stopped the work": Harlow Shapley and Helen E.
Howarth, eds., *A Sourcebook of Astronomy* (New York: McGraw-Hill,
1929), p. 321.

89 "The artist . . . in terms of masses": E. H. Gombrich, *Art and Illusion: A Study
in the Psychology of Pictorial Representation,* 2d ed. (Princeton, N.J.: Princeton University Press, 1961), p. 65.

90 "The network . . . beings": *Sourcebook of Astronomy,* p. 385.

90 "It seems that . . . this mystery": Flammarion, *Popular Astronomy,* p. 390.

90 "Could the meteorological . . . blade of grass?" Flammarion, *Popular Astronomy,* p. 385.

91 "enlarge the circle . . . every world": Flammarion, *Popular Astronomy,* p. 386.

91 "For some time . . . other world": Percival Lowell, *Mars* (Boston: Houghton,
Mifflin, 1895), p. 116.

92 "Beautiful as the . . . sinister intent": Percival Lowell, *Mars as the Abode of
Life* (New York: Macmillan, 1908), p. 134.

93 "winding, . . . details": E. M. Antoniadi, "Considerations on the Physical
Appearance of the Planet Mars," *Popular Astronomy* 21 (1913): 420.

94 "The sharpish . . . patterns we see," quoted in Patrick Moore, *Guide to Mars*
(New York: Norton, 1977), p. 90.

CHAPTER 8. ASTEROIDS

102 "I am having . . . factor": quoted in William Graves Hoyt, *Planets X and
Pluto* (Tucson: University of Arizona Press, 1980), p. 26.

105 "without any . . . time": quoted in Laurence G. Taff, *Celestial Mechanics*
(New York: John Wiley & Sons, 1985), p. 220.

105 "Just about this time . . . observations": Taff, *Celestial Mechanics,* p. 220.

105 "The methods first . . . work": quoted in Taff, *Celestial Mechanics,* p. 221.

106 "Did Ceres . . . catastrophe?": quoted in Ley, *Watchers,* p. 320.

112 "When Vesta . . . found": David A. Allen, letter, *Sky and Telescope* 67
(1984): 493.

CHAPTER 9. JUPITER

117 "That enormous . . . planet": quoted in Ley, *Watchers,* p. 343.

124 "in vain and too late": quoted in Suzanne Débarbat and Curtis Wilson,

"The Galilean Satellites of Jupiter from Galileo to Cassini, Römer, and Bradley," in *Planetary Astronomy from the Renaissance to the Rise of Astrophysics*. Part A: *Tycho Brahe to Newton*. ed. R. Taton and C. Wilson (Cambridge: Cambridge University Press, 1989), p. 146.

128 "Calculations . . . outgassing": S. J. Peale, P. Cassen, and R. T. Reynolds, "Melting of Io by Tidal Dissipation," *Science* 203 (1979): 892–894.

CHAPTER 10. SATURN

132 "I have discovered . . . one another": quoted in G. Abetti, *The History of Astronomy* (London: Sidgwick and Jackson, 1954), p. 104.

133 "by a thin . . . ecliptic": quoted in A. F. O'D. Alexander, *The Planet Saturn: A History of Observation, Theory, and Discovery* (1962; reprint, New York: Dover, 1980), p. 94.

133 "The mind . . . condition": quoted in Alexander, *The Planet Saturn*, p. 140.

135 "rigid . . . coherent": quoted in Stephen G. Brush, C.W.F. Everitt, and Elizabeth Garber, eds., *Maxwell on Saturn's Rings* (Cambridge, Mass.: MIT Press, 1983), p. 70.

137 "That the rings . . . confirmed": Percival Lowell, "Memoir on Saturn's Rings," *Memoirs of the Lowell Observatory* 1, no. 2 (1915): 3.

CHAPTER 11. URANUS

152 "The great . . . instruments": quoted in Constance A. Lubbock, *The Herschel Chronicle* (Cambridge: Cambridge University Press, 1933), p. 59.

154 "In the quartile . . . perhaps a comet": William Herschel, *The Scientific Papers*, ed. J.L.E. Dreyer (London: Royal Society and Royal Astronomical Society, 1912), p. xxix.

154 "looked for . . . place": Herschel, *Scientific Papers*, p. xxx.

154 "The diameter is . . . us": Herschel, *Scientific Papers*, p. xxx.

154 "as likely . . . ellipsis": Lubbock, *Herschel Chronicle*, p. 80.

154 "nothing . . . comet": Lubbock, *Herschel Chronicle*, p. 86.

155 "I cannot . . . His Auspicious Reign": quoted in Hoyt, *Planets X and Pluto*, p. 24.

156 "overcrowd our . . . devised": E. M. Antoniadi, "Report of Mars Section, 1896–97," *Memoirs of the British Astronomical Association* 5 (1897): 85.

CHAPTER 12. NEPTUNE

167 "never . . . observation": quoted in Edward S. Holden, *Sir William Herschel: His Life and Works* (New York: Scribner's, 1880), p. 5.

168 "I leave . . . planet": quoted in Morton Grosser, *The Discovery of Neptune* (1962; reprint, New York: Dover, 1979), p. 42.

168 "a subject . . . accurate sweep": George Biddell Airy, "Account of Some Circumstances Historically Connected with the Discovery of the Planet Exterior to Uranus," *Monthly Notices of the Royal Astronomical Society* 7 (1846): 123.

169 "It is . . . successive revolutions": Airy, "Account," 124.

170 "Formed . . . beyond it": quoted in Grosser, *Discovery of Neptune*, pp. 75–76.

170 "You see . . . I know": Grosser, *Discovery of Neptune*, p. 76.

171 "so novel . . . doubtful": quoted in Grosser, *Discovery of Neptune*, p. 135.

171 "clearly impossible . . . again": quoted in Grosser, *Discovery of Neptune*, p. 96.

172 "I think . . . process": Airy, "Account," p. 132.

172 "The difficulty . . . celestial mechanics": Agnes M. Clerke, *A Popular History of Astronomy During the Nineteenth Century*, 3d ed. (London: Adam and Charles Black, 1893), p. 98.

173 "a shoddy patchwork of errors": Grosser, *Discovery of Neptune*, p. 99.

173 "The radius . . . point": Airy, "Account," p. 134.

173 "You know . . . Uranus": quoted in Robert W. Smith, "The Cambridge Network in Action: The Discovery of Neptune." *Isis* 80 (1989): 410.

173 "In my opinion . . . delay": Airy, "Account," p. 136.

174 "I get over . . . year": Airy, "Account," p. 137.

174 "The author . . . never saw it": Flammarion, *Popular Astronomy*, p. 466.

174 "We see . . . demonstration": John Herschel, "Le Verrier's Planet," letter, *Athenaeum*, Oct. 3, 1846, p. 1019.

175 "to find . . . discover": quoted in Grosser, *Discovery of Neptune*, p. 115.

175 "The planet . . . *exists*": quoted in Grosser, *Discovery of Neptune*, p. 119.

176 "my strict duty . . . *Herschel*": quoted in Grosser, *Discovery of Neptune*, p. 125.

176 "not-so-subtle . . . proposal": Hoyt, *Planets X and Pluto*, p. 25.

176 "The remarkable . . . name": Herschel, "Le Verrier's Planet."

176 "Why has . . . telescope?": quoted in Grosser, *Discovery of Neptune*, p. 130.

177 "perfectly free from jealousy": quoted in Grosser, *Discovery of Neptune*, p. 143.

177 "much taste . . . trade": Baum, *The Planets*, p. 125.

177 "Look out . . . expedition!": this quotation, and quotations from the notebooks of William Lassell, were kindly provided by Richard Baum.

178 "intimately related to the telescope," quoted in Richard Baum and Robert W. Smith, "Neptune's Forgotten Ring," *Sky and Telescope* 77 (1989): 611.

178 "an ice-blue . . . windowpane": Stephen J. O'Meara, "Neptune Through the Eyepiece," *Sky and Telescope* 77 (1989): 487.

179 "Who can say . . . approach?": Webb, *Celestial Objects*, 1:272.

CHAPTER 13. PLUTO

186 "The planet Neptune . . . accident": quoted in Hoyt, *Planets X and Pluto*, p. 63.

188 "not only . . . yet known": Percival Lowell, *The Solar System* (Boston: Houghton, Mifflin, 1903), p. 17.

189 "there is . . . break": quoted in Hoyt, *Planets X and Pluto*, pp. 135–136.

189 "Owing to . . . rests": quoted in Hoyt, *Planets X and Pluto*, p. 140.

191 "It seemed . . . doing": quoted in Hoyt, *Planets X and Pluto*, p. 179.

191 "I ought . . . rascal": quoted in David H. Levy, *Clyde Tombaugh: Discoverer of Planet Pluto* (Tucson: University of Arizona Press, 1991), p. 46.

192 "I encountered . . . despair": Clyde Tombaugh and Patrick Moore, *Out of the Darkness: The Planet Pluto* (Harrisburg, Pa.: Stackpole Books, 1980), p. 117.

193 "I raised . . . to myself": Moore, *Out of the Darkness*, p. 126.

193 "I've found . . . evidence": Clyde Tombaugh, personal communication with the author.

193 "Systematic search . . . assigned": Hoyt, *Planets X and Pluto*, p. 196.

195 "I don't think . . . done": quoted in Hoyt, *Planets X and Pluto*, p. 222.

198 "seemed . . . apart": quoted in Stillman Drake and Charles T. Kowal, "Galileo's Sighting of Neptune." *Scientific American* 243 (1980): 78.

CHAPTER 14. COMETS AND METEORS

201 "write my name across the sky": quoted in John E. Bortle, "Comets and How to Hunt Them," *Sky and Telescope* 61 (1981): 125.

BIBLIOGRAPHY

GENERAL REFERENCES

Atreya, S. K., J. B. Pollack, and M. S. Matthews, eds. *Origin and Evolution of Planetary and Satellite Atmospheres.* Tucson: University of Arizona Press, 1989.

Baum, Richard. *The Planets: Some Myths and Realities.* Newton Abbot: David & Charles, 1973.

Beatty, J. Kelly, and Andrew Chaikin, eds. *The New Solar System.* 3d ed. Cambridge, Mass.: Sky Publishing Corp., 1990.

Burns, J. A., and M. S. Matthews, eds. *Satellites.* Tucson: University of Arizona Press, 1986.

Callatay, V. de, and A. Dollfus. *Atlas of the Planets.* London: Heinemann, 1974.

Carr, Michael H., et al. *The Geology of the Terrestrial Planets.* Washington, D.C.: NASA SP-469, 1984.

Chapman, Clark. *Planets of Rock and Ice.* New York: Charles Scribner's Sons, 1982.

Crowe, Michael J. *The Extraterrestrial Life Debate, 1750–1910: The Idea of a Plurality of Worlds from Kant to Lowell.* Cambridge: Cambridge University Press, 1986.

Dick, Steven J. *Plurality of Worlds: The Origins of the Extraterrestrial Life Debate from Democritus to Kant.* Cambridge: Cambridge University Press, 1982.

Doherty, Paul. *Atlas of the Planets.* New York: McGraw-Hill, 1980.

Greenberg, Richard, and Andre Brahic, eds. *Planetary Rings.* Tucson: University of Arizona Press, 1984.

Kuiper, G. P., and B. M. Middlehurst, eds. *Planets and Satellites.* Chicago: University of Chicago Press, 1961.

Ley, Willy. *Watchers of the Skies.* New York: Viking Press, 1963.

Moore, Patrick, and Garry Hunt. *Atlas of the Solar System.* London: Mitchell Beazley, 1983.

Murray, Bruce, Michael Malin, and Ronald Greeley. *Earthlike Planets*. San Francisco: W. H. Freeman, 1981.

Pannekoek, Anton. *A History of Astronomy*. 1961. Reprint. New York: Dover, 1989.

Sagan, Carl. *Cosmos*. New York: Random House, 1980.

Taton, René, and Curtis Wilson, eds. *Planetary Astronomy from the Renaissance to the Rise of Astrophysics*. Part A: *Tycho Brahe to Newton*. Cambridge: Cambridge University Press, 1989.

CHAPTER 1. WANDERING STARS

Caspar, Max. *Kepler*. London: Abelard-Schuman, 1959.

Christianson, Gale E. *This Wild Abyss: The Story of the Men Who Made Modern Astronomy*. New York: Free Press, 1978.

Copernicus, Nicholas. *On the Revolution of the Heavenly Spheres*. Trans. A. M. Duncan. Newton Abbot: David & Charles, 1976.

———. *Three Copernican Treatises*. Trans. Edward Rosen. 1939. Reprint. New York: Dover Publications, 1959.

Dreyer, J.L.E. *A History of Astronomy from Thales to Kepler*. 1906. Reprint. New York: Dover Publications, 1953.

Hoag, Arthur A. "Aristarchos Revisited." *Griffith Observer* 54 (1990): 10–18.

Koestler, Arthur. *The Sleepwalkers*. New York: Grosset & Dunlap, 1963.

Koyre, Alexandre. *The Astronomical Revolution*. Trans. R.E.W. Maddison. Ithaca, N.Y.: Cornell University Press, 1973.

Neugebauer, Otto. *The Exact Sciences in Antiquity*. 2d ed. 1957. Reprint. New York: Dover Publications, 1969.

Newton, Isaac. *Mathematical Principles of Natural Philosophy*. Trans. Andrew Motte (1729). Translation revised by Florian Cajori. Berkeley: University of California Press, 1934.

Van Helden, Albert. *Measuring the Universe*. Chicago: University of Chicago Press, 1985.

Westfall, Richard S. *Never at Rest: A Biography of Isaac Newton*. Cambridge: Cambridge University Press, 1980.

CHAPTER 2. THROUGH THE TELESCOPE

Drake, Stillman. *Galileo at Work: His Scientific Biography*. Chicago: University of Chicago Press, 1978.

Galilei, Galileo. *Dialogue Concerning the Two Chief World Systems: Ptolemaic and Copernican*. Trans. Stillman Drake. Berkeley: University of California Press, 1967.

———. *Discoveries and Opinions of Galileo*. Trans. and ed. Stillman Drake. Garden City, N.Y.: Doubleday, 1957.

———. *Sidereus Nuncius; or, The Sidereal Messenger*. Trans. Albert Van Helden. Chicago: University of Chicago Press, 1989.

Kepler, Johannes. *Kepler's Conversation with Galileo's Sidereal Messenger.* Trans. Edward Rosen. New York: Johnson Reprint Corp., 1965.

King, Henry C. *The History of the Telescope.* 1955. Reprint. New York: Dover Publications, 1979.

Newton, Isaac. *Opticks; or, A Treatise of the Reflections, Refractions, Inflections and Colours of Light.* New York: Dover Publications, 1982.

Santillana, Giorgio de. *The Crime of Galileo.* Chicago: University of Chicago Press, 1955.

Van Helden, Albert. *The Invention of the Telescope.* Philadelphia: American Philosophical Society, 1977.

CHAPTER 3. ROCKETS INTO SPACE

Armstrong, Neil, Michael Collins, and Edwin E. Aldrin. *First on the Moon.* Boston: Little, Brown, 1970.

Collins, Michael. *Carrying the Fire.* New York: Ballantine, 1974.

Kepler, Johannes. *Kepler's Somnium.* Trans. Edward Rosen. Madison, Wis.: University of Wisconsin Press, 1967.

Rycroft, Michael, ed. *The Cambridge Encyclopedia of Space.* Cambridge: Cambridge University Press, 1990.

CHAPTER 4. THE MOON

Baldwin, Ralph B. *The Face of the Moon.* Chicago: University of Chicago Press, 1949.

———. *The Measure of the Moon.* Chicago: University of Chicago Press, 1963.

Bowker, L.D.E., and J. K. Hughes. *Lunar Orbiter Photographic Atlas of the Moon.* Washington, D.C.: NASA SP-206, 1971.

Elger, T. Gwyn. *The Moon: A Full Description and Map of Its Principal Features.* London: George Philip & Son, 1895.

French, Bevan. *The Moon Book.* Penguin Books, 1977.

Goodacre, Walter. *The Moon.* Bournemouth, Eng.: Pardy & Son, 1931.

Hill, Harold. *A Portfolio of Lunar Drawings.* Cambridge: Cambridge University Press, 1991.

Hoyt, William Graves. *Coon Mountain Controversies: Meteor Crater and the Development of Impact Theory.* Tucson: University of Arizona Press, 1987.

Kopal, Zdenek, and R. W. Calder. *Mapping the Moon.* Dordrecht, Holland: D. Reidel Publishing Co., 1974.

Kosofsky, L. J., and F. El-Baz. *The Moon as Viewed by Lunar Orbiter.* Washington, D.C.: NASA SP-200, 1970.

Masursky, H., C. W. Colton, and Forouk El-Baz. *Apollo Over the Moon: A View from Orbit.* Washington, D.C.: NASA SP-362, 1978.

Whitaker, Ewen A. "Galileo's Lunar Observations and the Dating of the Composition of 'Sidereus Nuncius.'" *Journal for the History of Astronomy* 9 (1978): 155–169.

———. "Selenography in the Seventeenth Century." In *Planetary Astronomy from the Renaissance to the Rise of Astrophysics,* Part A: *Tycho Brahe to Newton,* ed. René Taton and Curtis Wilson. Cambridge: Cambridge University Press, 1989.

Wilhelms, Don E. *The Geologic History of the Moon.* Washington, D.C.: U.S. Government Printing Office, 1987.

CHAPTER 5. MERCURY

Antoniadi, E. M. *The Planet Mercury.* Trans. Patrick Moore. Devon, Eng.: Keith Reid, 1974.

Baum, Richard. "Leverrier and the Lost Planet." In *Yearbook of Astronomy 1982,* ed. Patrick Moore. New York: W. W. Norton, 1981.

Murray, Bruce C. "Mercury." *Scientific American* 233 (1975): 58–68.

Roseveare, N. T. *Mercury's Perihelion from Leverrier to Einstein.* Oxford: Clarenden Press, 1982.

Strom, Robert G. *Mercury: The Elusive Planet.* Washington, D.C.: Smithsonian Institution Press, 1987.

———. "Mercury: The Forgotten Planet." *Sky and Telescope* 80 (1990): 256–260.

Vilas, Faith, Clark R. Chapman, and Mildred S. Matthews, eds. *Mercury.* Tucson: University of Arizona Press, 1988.

CHAPTER 6. VENUS

Baum, Richard. "The Himalayas of Venus." In *The Planets: Some Myths and Realities,* 48–83. Newton Abbot: David & Charles, 1973.

Bazilevskiy, Aleksandr T. "The Planet Next Door." *Sky and Telescope* 77 (1989): 360–366.

Fimmel, R. O., L. Colin, and E. Burgess. *Pioneer Venus.* Washington, D.C.: NASA SP-46, 1983.

Gingerich, Owen. "Phases of Venus in 1610." *Journal for the History of Astronomy* 15 (1984): 209–210.

Hunten, D. M., L. Colin, T. M. Donahue, and V. I. Moroz, eds. *Venus.* Tucson: University of Arizona Press, 1983.

Kasting, James F., Owen B. Toon, and James B. Pollack. "How Climate Evolved on the Terrestrial Planets." *Scientific American* 258 (1988): 90–97.

Magellan at Venus. Science 252 (1991): 247–312. Special issue.

CHAPTER 7. MARS

Antoniadi, E. M. *The Planet Mars.* Trans. Patrick Moore. Newton Abbot: David & Charles, 1973.

Carr, Michael H., et al. *Viking Orbiter Views of Mars.* Washington, D.C.: NASA SP-441, 1980.

———. *The Surface of Mars.* New Haven: Yale University Press, 1981.

Dick, Steven J. "Discovering the Moons of Mars." *Sky and Telescope* 76 (1988): 242–243.

Haberle, R. M. "The Climate of Mars." *Scientific American* 254 (1986): 54–62.

Hoyt, William Graves. *Lowell and Mars.* Tucson: University of Arizona Press, 1976.

Lowell, Percival. *Mars.* Boston: Houghton, Mifflin, 1895.

———. *Mars and Its Canals.* New York: Macmillan, 1908.

———. *Mars as the Abode of Life.* New York: Macmillan, 1908.

Mutch, Thomas A., et al. *The Geology of Mars.* Princeton, N.J.: Princeton University Press, 1976.

Sheehan, William. *Planets and Perception: Telescopic Views and Interpretations, 1609–1909.* Tucson: University of Arizona Press, 1988.

Viking Lander Imaging Team. *The Martian Landscape.* Washington, D.C.: NASA SP-425, 1978.

CHAPTER 8. ASTEROIDS

Alvarez, Luis W., Walter Alvarez, Frank Asaro, and Helen V. Michel. "Extraterrestrial Cause for the Cretaceous-Tertiary Extinction." *Science* 208 (1980): 1095–1108.

Beatty, J. Kelly. "Killer Crater in the Yucatán?" *Sky and Telescope* 82 (1991): 38–40.

Binzel, Richard P., Tom Gehrels, and Mildred S. Mathews, eds. *Asteroids II.* Tucson: University of Arizona Press, 1989.

Cunningham, Clifford J. *Introduction to Asteroids.* Richmond, Va.: Willmann-Bell, 1988.

Gehrels, Tom, ed. *Asteroids.* Tucson: University of Arizona Press, 1979.

Kowal, Charles T. *Asteroids: Their Nature and Utilization.* New York: John Wiley & Sons, 1988.

Morrison, David, and Clark R. Chapman. "Target Earth: It Will Happen." *Sky and Telescope* 79 (1990): 261–265.

CHAPTER 9. JUPITER

Beebe, Reta F. "Queen of the Giant Storms." *Sky and Telescope* 80 (1990): 359–364.

Gehrels, Tom, ed. *Jupiter.* Tucson: University of Arizona Press, 1976.

Ingersoll, Andrew P. "Jupiter and Saturn." *Scientific American* 245 (1981): 66–80.

Morrison, David, and Jane Samz. *Voyage to Jupiter.* Washington, D.C.: NASA SP-439, 1980.

Morrison, David, ed. *Satellites of Jupiter.* Tucson: University of Arizona Press, 1982.

Peale, S. J., P. Cassen, and R. T. Reynolds. "Melting of Io by Tidal Dissipation." *Science* 203 (1979): 892–894.

Peek, B. M. *The Planet Jupiter*. London: Faber and Faber, 1958.

Soderblom, Laurence A. "The Galilean Moons of Jupiter." *Scientific American* 242 (1980): 88–100.

CHAPTER 10. SATURN

Alexander, A. F. O'D. *The Planet Saturn: A History of Observation, Theory, and Discovery*. 1962. Reprint. New York: Dover Publications, 1980.

Brush, Stephen G., C.W.F. Everitt, and Elizabeth Garber, eds. *Maxwell on Saturn's Rings*. Cambridge, Mass.: MIT Press, 1983.

Cuzzi, Jeffrey N. "Ringed Planets: Still Mysterious." *Sky and Telescope* 68 (1984): 511–516, and 69 (1985): 19–23.

Gehrels, Tom, and Mildred S. Matthews, eds.. *Saturn*. Tucson: University of Arizona Press, 1984.

Morrison, Donald. *Voyages to Saturn*. Washington, D.C.: NASA SP-451, 1982.

O'Meara, Stephen J. "Saturn's Great White Spot Spectacular." *Sky and Telescope* 81 (1991): 144–147.

Owen, Tobias. "Titan." *Scientific American* 246 (1982): 76–93.

Pollack, James B., and Jeffrey N. Cuzzi. "Rings in the Solar System." *Scientific American* 245 (1981): 78–93.

Sanchez-Lavega, Agustin. "Saturn's Great White Spots." *Sky and Telescope* 78 (1989): 141–142.

Soderblom, Laurence A., and Torrence V. Johnson. "The Moons of Saturn." *Scientific American* 246 (1982): 73–86.

Van Helden, Albert. "Saturn and His Anses." *Journal for the History of Astronomy* 5 (1974): 105–121.

CHAPTER 11. URANUS

Alexander, A. F. O'D. *The Planet Uranus: A History of Observation, Theory, and Discovery*. London: Faber & Faber, 1965.

Baum, Richard. "William Herschel and the Rings of Uranus." In *The Planets: Some Myths and Realities*, 106–119. Newton Abbot: David & Charles, 1973.

Beatty, J. Kelly. "A Place Called Uranus." *Sky and Telescope* 7 (1986): 333–337.

Bergstralh, Jay T., Ellis D. Miner, and Mildred Shapley Matthews, eds. *Uranus*. Tucson: University of Arizona Press, 1991.

Chaikin, Andrew. "Voyager Among the Ice Worlds." *Sky and Telescope* 71 (1986): 338–343.

Cuzzi, Jeffrey N., and L. W. Esposito. "The Rings of Uranus." *Scientific American* 257 (1987): 52–66.

Elliot, James, et al. "Discovering the Rings of Uranus." *Sky and Telescope* 53 (1977): 412–416.

Elliot, James, and Richard Kerr. *Rings: Discoveries from Galileo to Voyager*. Cambridge, Mass.: MIT Press, 1984.

Goldreich, Peter, and Scott Tremaine. "Towards a Theory for the Uranian Rings." *Nature* 277 (1979): 97–99.

Hoskin, Michael A. *William Herschel: Pioneer of Sidereal Astronomy.* London: Sheed and Ward, 1959.

Hunt, Garry, ed. *Uranus and the Outer Planets.* Cambridge: Cambridge University Press, 1982.

Ingersoll, Andrew P. "Uranus." *Scientific American* 255 (1987): 38–45.

Johnson, Torrence V., Robert Hamilton Brown, and Lawrence A. Soderblom. "The Moons of Uranus." *Scientific American* 256 (1987): 48–60.

Littman, Mark. *Planets Beyond: Discovering the Outer Solar System.* 2d ed. New York: John Wiley, 1990.

Lubbock, Constance A. *The Herschel Chronicle.* Cambridge: Cambridge University Press, 1933.

O'Meara, Stephen J. "A Visual History of Uranus." *Sky and Telescope* 70 (1985): 411–412.

CHAPTER 12. NEPTUNE

Baum, Richard, and Robert W. Smith. "Neptune's Forgotten Ring." *Sky and Telescope* 77 (1989): 610–611.

Grosser, Morton. *The Discovery of Neptune.* 1962. Reprint. New York: Dover Publications, 1979.

Moore, Patrick. *The Planet Neptune.* New York: John Wiley, 1988.

Smith, Robert W. "The Cambridge Network in Action: The Discovery of Neptune." *Isis* 80 (1989): 395–422.

Voyager 2 at Neptune. Science 246 (1989): 1417–1453. Special issue.

CHAPTER 13. PLUTO

Beatty, J. Kelly, and Anita Killian. "Discovering Pluto's Atmosphere." *Sky and Telescope* 76 (1988): 624–627.

Drake, Stillman, and Charles T. Kowal. "Galileo's Sighting of Neptune." *Scientific American* 243 (1980): 74–79.

Hoyt, William Graves. *Planets X and Pluto.* Tucson: University of Arizona Press, 1980.

Levy, David H. *Clyde Tombaugh: Discoverer of Planet Pluto.* Tucson: University of Arizona Press, 1991.

Littman, Mark. "Where is Planet X?" *Sky and Telescope* 78 (1989): 596–599.

Tombaugh, Clyde, and Patrick Moore. *Out of the Darkness: The Planet Pluto.* Harrisburg, Pa.: Stackpole Books, 1980.

CHAPTER 14. COMETS AND METEORS

Brandt, J. C., and R. D. Chapman. *Introduction to Comets.* Cambridge, Eng.: Cambridge University Press, 1981.

Dodd, R. T. *Thunderstones and Shooting Stars.* Cambridge, Mass.: Harvard University Press, 1986.

Kerridge, J. F., and M. S. Matthews, eds. *Meteorites and the Early Solar System.* Tucson: University of Arizona Press, 1988.

Sagan, Carl, and Ann Druyan. *Comet.* New York: Random House, 1985.

Weissman, Paul R. "Realm of the Comets." *Sky and Telescope* 73 (1987): 238–241.

Whipple, Fred W. *The Mystery of Comets.* Washington, D.C.: Smithsonian Institution Press, 1985.

———. "The Black Heart of Halley's Comet." *Sky and Telescope* 73 (1987): 242–245.

Wilkening, Laurel L., ed. *Comets.* Tucson: University of Arizona Press, 1982.

INDEX

Page numbers in italics indicate illustrations.

ABOUT THE AUTHOR

WILLIAM SHEEHAN is an amateur astronomer, psychiatrist, and writer. He received a bachelor's degree from the University of Minnesota, a master's degree from the University of Chicago, and an M.D. degree from the University of Minnesota. He also completed residency training in psychiatry at the University of Minnesota. Sheehan's first book, *Planets and Perception: Telescopic Views and Interpretations, 1609–1909*, was named a 1988 book of the year by the Astronomical Society of the Pacific. Dr. Sheehan has also published numerous articles in astronomy and psychiatry. His interest in the Moon and planets goes back to the early days of the spacecraft era, and he regularly observes with a 6-inch refractor and a 12 ½-inch reflector from his home in St. Paul, Minnesota.